NUCLEAR MATERIALS AND DISASTER RESEARCH

REMEDIATION OF LANDS AFTER THE FUKUSHIMA DAIICHI ACCIDENT

NUCLEAR MATERIALS AND DISASTER RESEARCH

Additional books in this series can be found on Nova's website under the Series tab.

Additional E-books in this series can be found on Nova's website under the E-book tab.

NUCLEAR MATERIALS AND DISASTER RESEARCH

REMEDIATION OF LANDS AFTER THE FUKUSHIMA DAIICHI ACCIDENT

HELEN G. TURNER
EDITOR

nova publishers
New York

For permission to use material from this book please contact us:
Telephone 631-231-7269; Fax 631-231-8175
Web Site: http://www.novapublishers.com

Library of Congress Cataloging-in-Publication Data

ISBN: 978-1-62948-846-2

Published by Nova Science Publishers, Inc. † New York

CONTENTS

PREFACE

The Fukushima Nuclear Power Station Accident in March 2011 significantly impacted Japan and the Japanese People, contaminating large areas of Fukushima Prefecture with radiation. Japan's nuclear incident has engendered much public and congressional concern about the possible impact of radiation on the Japanese public, as well as possible fallout on U.S. citizens. This book provides observation and commentary on remediation of the lands off-site from the Fukushima Daiichi reactors, and information on technical aspects of the nuclear incident, with reference to human health.

Chapter 1 – The United States (U.S.) Department of State, in conjunction with the U.S. Department of Energy (DOE) and the U.S. Environmental Protection Agency (EPA), sponsored three Embassy Science Fellows to provide expertise to support the Government of Japan (GOJ) Ministry of the Environment (MOE) in executing its charter under the GOJ Act on Special Measures to sponsor and oversee decontamination work off-site from the Fukushima Daiichi nuclear power plant (NPP).

This report provides the Embassy Science Fellows' observations and recommendations about the decontamination work and progress in the environmental remediation of the lands contaminated by the Daiichi NPP accident.

During their two-month mission in Japan, the ESFs met with MOE staff, staff of other agencies directly supporting the MOE in the decontamination work, and contractors performing the actual decontamination work; reviewed their plans and reports describing their work and results to date; and toured facilities and decontamination sites. The ESF team received excellent cooperation from all agencies and decontamination contractors. Particularly noteworthy was the strong desire expressed by all to reclaim the lands for renewed prosperity in the region.

Chapter 2 – Japan's nuclear incident has engendered much public and congressional concern about the possible impact of radiation on the Japanese public, as well as possible fallout on U.S. citizens. This report provides information on technical aspects of the nuclear incident, with reference to human health.

While some radioactive material from the Japanese incident may reach the United States, it appears most unlikely that this material will result in harmful levels of radiation. In traveling thousands of miles between the two countries, some radioactive material will decay, rain will wash some out of the air, and its concentration will diminish as it disperses.

Many atoms are stable; they remain in their current form indefinitely. Other atoms are unstable, or radioactive. They "decay" or "disintegrate," emitting energy through various forms of radiation. Each form has its own characteristics and potential for human health effects.

Nuclear reactors use uranium or mixed oxides (uranium oxide and plutonium oxide, or MOX) for fuel. Uranium and plutonium atoms fission, or split, releasing neutrons that cause additional fissions in a chain reaction, and also releasing energy. A nuclear reactor's core consists of fuel rods made of uranium or MOX encased in zirconium, and neutron-absorbing control rods that are removed or inserted to start or stop the chain reaction. This assembly is placed underwater to carry off excess heat. The incident at the Fukushima Daiichi Nuclear Power Plant prevented water from circulating in the core of several reactors, causing water to evaporate and temperature to rise. High heat could melt the fuel rods and lead to a release of radioactive material into the air.

When uranium and plutonium fission, they split into smaller atoms that are highly radioactive and generate much heat; indeed, fuel rods that have just been removed from a reactor are much more radioactive, and hotter, than fuel rods before they have been inserted into a reactor. After fuel rods can no longer efficiently produce energy, they are considered "spent" and are placed in cooling pools of water for several years to keep them from overheating while the most radioactive materials decay. A concern about the spent fuel pool at reactor 4 is that it may have lost most or all of its water, yet it has more fuel rods than pools at the other five reactors, as it contains all the active fuel rods that were temporarily removed from the reactor core in November 2010 to permit plant maintenance in addition to spent fuel rods.

A nuclear reactor cannot explode like an atomic bomb because the concentration of the type of uranium or plutonium that fissions easily is too low to support a runaway chain reaction, and a nuclear weapon requires one of two configurations, neither of which is present in a reactor.

Some types of radiation have enough energy to knock electrons off atoms, creating "ions" that are electrically charged and highly reactive. Ionizing radiation is thus harmful to living cells. It strikes people constantly, but in doses low enough to have negligible effect. A concern about the reactor incident is that it will release radioactive materials that pose a danger to human health. For example, cesium-137 emits gamma rays powerful enough to penetrate the body and damage cells. Ingesting iodine-131 increases the risk of thyroid cancer. Potassium iodide tablets protect the thyroid, but there is no need to take them absent an expectation of ingesting iodine-131.

In: Remediation of Lands After the Fukushima ... ISBN: 978-1-62948-846-2
Editor: Helen G. Turner

Chapter 1

OBSERVATIONS AND COMMENTARY ON REMEDIATION OF THE LANDS OFF-SITE FROM THE FUKUSHIMA DAIICHI REACTORS*

Sang Don Lee, Robert L. Sindelar and Mark B. Triplett

FOREWORD

The Fukushima Nuclear Power Station Accident in March 2011 significantly impacted Japan and the Japanese People, contaminating large areas of Fukushima Prefecture with radiation. Cesium contamination is of particular concern to the Government and People of Japan, considering its relatively long half-life and dangers to human health. The Government of Japan (GOJ) and the United States Government (USG) have engaged in significant technical collaborations and partnerships related to decommissioning and decontamination following the March 2011 accident. We have worked closely together in support of GOJ efforts to decontaminate large areas of land as rapidly as possible, allowing residents of Fukushima Prefecture to move back into their homes and resume their normal lives.

In January 2013, as part of our continuing commitment to support the restoration of Fukushima Prefecture, the USG dispatched three

* This is an edited, reformatted and augmented version of the United States Embassy Fellows Support to the Government of Japan--Ministry of the Environment (Mission Dates: 4 February-29 March 2013), dated July 2013.

decontamination subject matter experts to support the Japanese Ministry of the Environment's (MOE) decontamination efforts in urban and residential areas. One subject matter expert from the Environmental Protection Agency (EPA) and two Department of Energy (DOE) national laboratory experts from Pacific Northwest National Laboratory and Savannah River National Laboratory were deployed under the U.S. Department of State's Embassy Science Fellows Program. This is the first time the USG has deployed decontamination subject matter experts in the Embassy Science Fellows Program. For a period of more than two months, which lasted until the end of March 2013, these U.S. decontamination subject matter experts worked intensively with MOE personnel to examine MOE's operations, share U.S. experiences, and offer suggestions for enhancing decontamination efforts from a systems integration standpoint.

For the duration of our Embassy Science Fellows partnership, the MOE provided significant amounts of information on its decontamination efforts in an open and transparent manner. The MOE also organized site tours in Fukushima Prefecture, and facilitated meetings with MOE subject matter experts on core aspects of MOE's restoration operations involving the National Institute of Environmental Studies, the Japan Atomic Energy Agency, Fukushima Prefecture and its municipalities, and decontamination contractors. Our partnership was successful due to MOE's excellent cooperation.

The recommendations contained in this document are the result of significant effort by the three U.S. experts to assess MOE's current, complex program of work and to suggest enhancements to current decontamination operations. We truly hope that these recommendations will be useful to the MOE and more broadly to the GOJ in connection with continuing and future efforts on the restoration of Fukushima Prefecture. Following the Embassy Science Fellows Program, we remain committed to supporting GOJ efforts on the restoration of Fukushima Prefecture and we look forward to continuing our partnership in the U.S.- Japan Bilateral Commission on Civil Nuclear Energy's Fukushima Working Group—an interagency group that works collaboratively to address technical challenges in the areas of decommissioning and decontamination.

In closing, I would like to underline our gratitude to the MOE as well as to the United States Department of State, the United States Department of Energy, the United States Environmental Protection Agency, the United States Department of Commerce, Pacific Northwest National Laboratory and

Savannah River National Laboratory for their hard work and dedication to make this bilateral partnership a success.

Jeffrey A. Miller
Energy Attaché
Director, U.S. Department of Energy Japan Office
U.S. Embassy Tokyo

ACRONYMS

BWR	Boiling Water Reactor
CAB	Citizens Advisory Board
DOE	US Department of Energy
EPA	US Environmental Protection Agency
ESF	Embassy Science Fellow
GOJ	Government of Japan
IAEA	International Atomic Energy Agency
ICRP	International Commission on Radiation Protection
ICS	Intensive Contamination Survey
JAEA	Japan Atomic Energy Agency
JNES	Japan Nuclear Energy Safety Organization
MAFF	Ministry of Agriculture, Forestry and Fisheries
METI	Ministry of Economy, Trade and Industry
MEXT	Ministry of Education, Culture, Sports, Science and Technology
MHLW	Ministry of Health, Labour and Welfare
MOE	Ministry of the Environment
NIES	National Institute for Environmental Studies
NISA	Nuclear Industrial Safety Agency
NPP	Nuclear Power Plant
NRA	Nuclear Regulation Authority
NSC	Nuclear Safety Commission
PNNL	Pacific Northwest National Laboratory
RA	Japan Reconstruction Agency
SDA	Special Decontamination Area
SRNL	Savannah River National Laboratory
TEPCO	Tokyo Electric Power Company
USJF	US Joint Force

EXECUTIVE SUMMARY

The United States (U.S.) Department of State, in conjunction with the U.S. Department of Energy (DOE) and the U.S. Environmental Protection Agency (EPA), sponsored three Embassy Science Fellows to provide expertise to support the Government of Japan (GOJ) Ministry of the Environment (MOE) in executing its charter under the GOJ Act on Special Measures to sponsor and oversee decontamination work off-site from the Fukushima Daiichi nuclear power plant (NPP).

This report provides the Embassy Science Fellows' observations and recommendations about the decontamination work and progress in the environmental remediation of the lands contaminated by the Daiichi NPP accident.

During their two-month mission in Japan, the ESFs met with MOE staff, staff of other agencies directly supporting the MOE in the decontamination work, and contractors performing the actual decontamination work; reviewed their plans and reports describing their work and results to date; and toured facilities and decontamination sites. The ESF team received excellent cooperation from all agencies and decontamination contractors. Particularly noteworthy was the strong desire expressed by all to reclaim the lands for renewed prosperity in the region.

The ESF team observed that the MOE was adapting to the unprecedented challenges of its charter and vigorously pursuing remediation of the contaminated lands to expedite return of the impacted area to the people, including

- rapidly mobilizing resources to respond to the unprecedented situation
- seeking, developing, testing, and applying innovative decontamination methods
- enabling local communities to have strong roles in forming decontamination plans and in siting waste storage facilities
- providing public education materials regarding radiation risks and status of decontamination efforts
- seeking to be open with the prefectural and municipal governments and citizens regarding decontamination activities
- engaging the international community to find the best solutions for remediation.

To accomplish their mission, the authors of this report drew upon their U.S. experience in large-scale remediation in DOE and EPA projects to identify the elements of remediation and highlighted strengths and suggested areas for improvement in the decontamination work. The ESFs provided U.S. information, lessons learned, and potential technologies that might benefit the remediation activities in Japan.

The ESF team developed the framework of program elements for environmental remediation of a population region contaminated by cesium from a total system perspective, as shown in the figure below. Information gathered and evaluated by the team was categorized into each of these elements.

Detailed recommendations for improvements to the remediation system elements are summarized in Table 1 below and discussed in Chapters 2 through 8. Implementation of the recommendations is expected to effect improved efficiency and cost-effectiveness of the remediation.

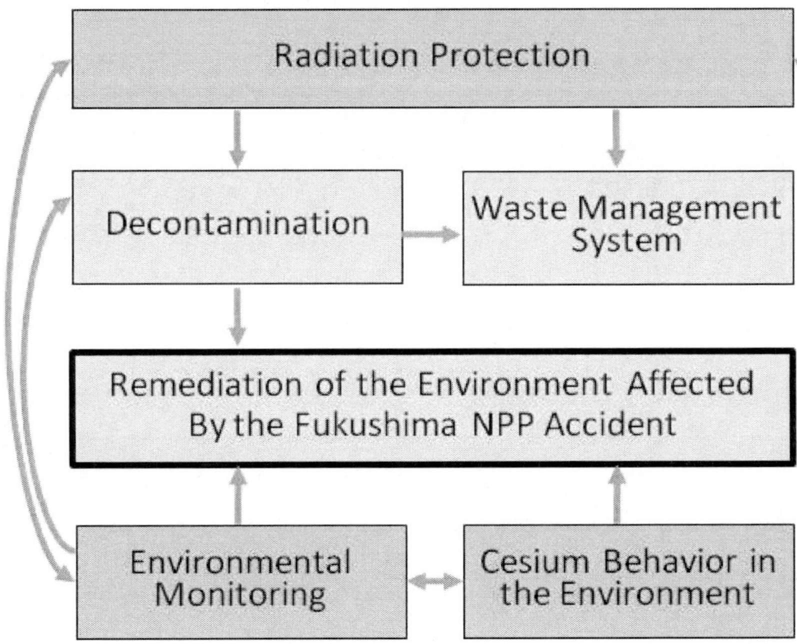

Program Elements for an Environmental Remediation System for a Populated Region Contaminated by Cesium.

Table 1. Recommendations for Improvements in the GOJ Remediation System

Radiation Protection		
Recommendations	**Rationale/Benefit**	**Supporting Actions**
1. Develop repopulation and dose reduction framework and implementing process for application at a community specific level.	• Provides guidance to local communities to support re-population decisions. • Provides defensibility of radiation protection standards based on international standards and practices. • Provides flexibility to accommodate local conditions and situations.	• Establish Expert Advisory Group on radiation protection to support and interact with prefectural and municipal government officials, and to provide necessary information to the public and stakeholder groups. • Develop guidance for remediation end-states (e.g., “how clean is clean?” or remediation completion criteria) and re-population standards and approaches that recognizes the unique characteristics of the incident and local considerations. Consider use of a phased approach for re-population efforts where the critical service providing personnel who return to the evacuated areas (e.g., public service providers, firemen, police, etc.) wear dosimeters to validate the actual dose exposure conditions in the community. • Develop a dose estimation process that accounts for specific land uses and population groups (e.g., industrial, residential, schools, etc.). For example using a tool such as RESRAD along with actual dosimetry data would provide a more useful estimate of potential dose to population groups. • Include community/stakeholder involvement process (formally chartered working group) to ensure that local considerations are incorporated into radiation protection processes. • Identify institutional controls (e.g., avoidance of “hot spots,” restrictions on access to forest areas that have not been decontaminated, etc.) that support ALARA principles for managing individual dose rates.

Radiation Protection		
Recommendations	**Rationale/Benefit**	**Supporting Actions**
2. Establish a radiation dosimetry program for residents who return to evacuated areas to provide the best information possible for understanding and managing population radiation exposure.	• Builds upon the dosimetry program that has been implemented within Fukushima Prefecture for municipal areas that were not subject to evacuation. • Provides a data set for validating dose prediction models and assumptions. • Provides an important data set for reviewing and potentially adapting re-population guidelines based on actual conditions and experience. • Provides a data set for long-term studies of potential health effects.	• Provide Expert Advisory Group and community/stakeholder interest groups to design and review the dosimetry program for returning population in the evacuation required areas. These groups should work collectively with government decision makers (national, prefectural, and community) to provide advice and to work collectively to resolve differences. • Implement dosimetry program and regularly collect data for review (item #3). • Conduct a systematic review of existing decontamination, health survey data, monitoring information including dosimetry data from affected population groups in Fukushima Prefecture. There are data collected from the previous dosimetry program in multiple cities in Fukushima Prefecture in 2011 and 2012.
3. Regularly review environ-mental monitoring results, dosimetry results and impacts from decontamination efforts to adapt the framework in item #1.	• Ensures that the re-population and dose reduction guidelines are effective and modified based on actual conditions and experience.	• The repopulation and dose reduction guidelines should be revisited as progress in remediation is achieved and as progress in community restoration occurs. • The review process should include the Expert Advisory Group (item #4) and community/stakeholder interest groups.
4. Establish Expert Advisory Group on radiation protection to provide technical assistance to prefectural and municipal government officials, and to provide necessary information to the public and stakeholder groups.	• Provides residents and government officials (community, prefectural and national) with the best available scientific, technical and medical expertise and advice.	• The independent Expert Advisory Group should include selec-ted National, Local, and private sector subject matter experts in fields as health physics, radiation protection, cost and risk analysis, remediation protection, and relevant regulatory requir-ements. This group periodically provides multi-disciplinary expert input on the radiation protection guidelines including technical issues, analysis of data, analysis of regulatory requirements, cost analyses and risk analyses (items #1 - #3)

Table 1. (Continued)

Decontamination Methods		
Recommendations	**Rationale/Benefit**	**Supporting Actions**
1. Develop and ensure application of a set of standard protocols for measuring the effectiveness of decontamination methods (before-after) for all applicable targets of decontamination (e.g., roads, soil, etc.).	• Ensures that the dose reduction impact of specific methods applied to specific surfaces is clearly understood, and distinct from overall dose reduction impacts from surrounding surfaces. Tells "how well" a particular decontamination method works. • Provides an essential base of information for defining areas needing improvement.	• Provide guidance for consistent and systematic application of the protocols (item #1) to the decontamination efforts that are underway in: (a) local communities, (b) MOE-led decontamination activities in the evacuation zones, and (c) MOE-led efforts in high-dose areas. • Provide contractual guidance to require implementation of these protocols and reporting of relevant data. • Collect and organize decontamination performance data that are generated in item #1 above in a manner that supports systematic analysis of the performance of the methods used for the full range of surfaces and land conditions.
2. Conduct a systematic anal-ysis of the existing performa nce data to identify potential factors or practices that could improve effectiv-eness of future decontamination efforts and that identifies situations where specific practices are not likely to be effective.	• Sets priorities and requirements for development, testing and deployment of improved decontamination methods.	• Identify problematic surfaces or material types that can be used to identify priorities in item #4. • Provide technology performance data to support item #3 (decontamination technology catalog).
3. Develop and maintain a comprehensive catalog of decontamination technology performance (based on systematic methods for assessing effectiveness).	• This catalog will define the "best practices" in the application of decontamination methods.	• Include key performance characteristics related to implementing each method including effectiveness, cost, materials, waste generation, etc.

Decontamination Methods		
Recommendations	**Rationale/Benefit**	**Supporting Actions**
4. Enhance existing processes for facilitating the development and maturing advanced decontamination technologies.	• Ensures that a sense of "continuous improvement" is maintained. • Provides continued attention on the more problematic decontamination challenges.	• Develop a government-provided test bed for advanced decontamination technology for the more intractable decontamination challenges such as forest lands and agricultural lands. The test-bed provides a readily available area with representative contamination conditions that can be available to companies, universities, laboratories, or others offering innovative or experimental methods. • Include contract incentives for soliciting advanced decontamination technologies. • Use the to-be established field decontamination effectiveness protocol in item #1 to evaluate new technologies.
Waste Management System[1]		
1. Expedite implementation of Temporary Storage Facilities (TSFs) in Intensive Decontamination Survey Areas and in Special Decontamination Areas.	• Alleviate accumulation of decontamination waste at distributed accumulation points (non-compliant storage). • Promptly reduce the hazard from distributed or in- place storage. • Show real progress.	• Encourage selection of TSFs prior to large scale decontamination work; one possible pre-condition for proceeding with decontamination work in any given community can be that community's selection of a fully compliant TSF to receive the material. • Increase communication with local governments and showcase successful TSFs. Consider empowering community/stakeholder working groups to provide advice on the holistic remediation AND waste management approaches for each community. • Consider use of nationally-owned or publicly-owned land (e.g., contaminated forest land).
2. Develop a waste inventory forecasting and tracking capability that incorporates a systems approach.	• Planning, evaluation and optimization of future storage, treatment and disposal approaches and facilities must be based on credible estimates of the volumes	• Provide a template for collecting a comprehensive data set from each municipality: volume by type of waste (soil, combustible, non-combustible, etc.); location or facility type; surface dose rates for bags or containers; and activity concentrations (Bq/kg). • Use the forecasting tool to conduct options analyses to support

Table 1. (Continued)

Waste Management System[1]		
Recommendations	**Rationale/Benefit**	**Supporting Actions**
	• and characteristics of waste materials to be generated. • Provides a credible set of estimates of the volumes and characteristics of decontamination wastes that are being generated. • This information is essential to planning and optimizing the treatment, storage, and disposal system components.	• analysis of high-level remediation strategies and priorities, including assessment of treatment, storage and transportation options and impacts.
3. Promptly implement modular, expandable Interim Storage Facilities (ISFs).	• Progress toward establishing ISFs could alleviate an obstacle to selecting TSF locations. • A modular/expandable design will allow optimization of volume reduction or other treatment processes.	• Complete surveys/investigations of candidate sites in Futaba, Okuma and Naraha towns. • Develop design concepts that balance the need to start receiving waste as soon as possible (e.g., by the target date of January 2015) but also allow evaluation of treatment methods (e.g., volume reduction and/or stabilization) that could reduce total storage requirements and produce more robust waste forms for final future disposal. • Design ISF to have expandable modules compatible with the land topography of the selected locations. • Design ISF to have areas for testing, demonstration and implementation of treatment methods (e.g., volume reduction). • Accommodate treatment decision from Recommendation #4. • Complete the facility design and prepare an evaluation of ISF options to allow for public review and comment on the ISF site selection and technical approach.

Waste Management System[1]		
Recommendations	**Rationale/Benefit**	**Supporting Actions**
		• Update the ISF implementation schedule based on an assessment of the current status and remaining actions for deployment.
4. Conduct systematic evaluation of treatment options for stabilization and/or volume reduction of decontamination waste.	• Enable early start of ISF but allow optimization of the life-cycle treatment, storage and disposal system. • Provide an explicit decision process for evaluating and selecting treatment approaches (pre-empt the frequent un-solicited proposals from private vendors).	• Identify candidate concepts for testing and implementing large scale treatment systems (e.g., volume reduction) for decontamination waste. • Provide a systematic technical, cost, and benefit evaluation of candidate treatment options that reflect a full range of potential options (no action, incineration, soil washing with reuse, thermal methods and distributed versus centralized methods). • Publish results with preferred alternative for public review and comment. • Place contracts to design, build and operate and consider subsequent step to evaluate treatment options.
5. Develop final disposal standards and regulations for decontamination waste.	• By understanding the end-state requirements for decontamination waste, more rational storage and treatment systems can be designed. • This is an essential step before evaluating and selecting a final disposal site.	• Evaluate estimated characteristics of decontamination waste categories relative to existing Japanese and international radioactive waste regulations or guidelines. • If necessary, develop disposal system and waste form requirements based on a performance assessment using acceptable international standards and methods. • For selected waste categories (e.g., contaminated soil, trees, etc.) evaluate the potential costs and benefits of coordinating some elements of off-site disposal with Daiichi decommissioning and remediation waste materials generated on site.
Environmental Monitoring		
1. Develop and implement an overall environmental	• Ensures that the requirements and methods for data collection are	• This plan needs to address fundamental QA/QC requirements for all monitoring data. Also the end use or purpose of data

Table 1. (Continued)

Environmental Monitoring		
Recommendations	**Rationale/Benefit**	**Supporting Actions**
monitoring plan that strengthens the linkage between the purpose/need for data and the data collection and management protocols.	driven by the need for and use of the data for decision making. • Ensures that data quality is controlled and maintained. • Provides confidence that monitoring data can be used for subsequent interpretation, analysis and decision making. • Provides central direction and consistent guidance for all relevant Fukushima environmental monitoring efforts.	collection needs to be clearly specified to ensure that data collection methods, resolution of data, frequency of collection, etc. meet overall needs. • This plan should address: 1. purpose of the data, 2. predefined data quality objective (what kind of resolution, accuracy, precision, duration, monitoring/sampling interval, etc.), 3. chain of custody (basic information of data, who collected it and where, record of data transfer), 4. quality assurance/quality control (each data set needs to go through predefined QA/QC protocol), 5. reporting (how the data will be reported), 6. Storage (how and where the data will be stored and available).
2. Enhance the data management systems to improve the consistency of data storage methods and accessibility to facilitate visualization and multi-disciplinary data evaluation and analysis	• Provides an efficient means for technical analysts to access and evaluate multiple data sets. • Facilitates visualization of multiple data sets • Ensures that there are common standards in place for communicating and accessing relevant monitoring data.	• Provide implementing guidelines for multiple agencies that are collecting monitoring data. • Nuclear Regulation Authority to develop a web-based data integration function.
3. Conduct periodic reviews and evaluations of monitoring data to ensure appropriate feedback with other strategic functions including efforts to	• Ensures that monitoring data are used to improve approaches for radiation protection and remediation. • Provides essential feedback to	• Maintain linkage to radiation protection to ensure that radiation • protection strategies are adapted to the most current and relevant conditions. • Maintain linkage to remediation strategy so that priorities for action can be focused on the most important aspects of

Environmental Monitoring		
Recommendations	**Rationale/Benefit**	**Supporting Actions**
optimize decontamination strategies, efforts to improve understanding of cesium behavior in the environment, and efforts to optimize the long-term monitoring program.	cesium fate and transport modeling to support model validation and improvement. • Provides process to quickly recognize anomalous or hazardous situations for which new actions may be needed.	remediation. • Maintain linkage to cesium behavior in the environment so that models can be validated and anomalous monitoring observations can be investigated.
Cesium Behavior in the Environment		
1. Continue development of cesium fate and transport models to enhance the ability to predict cesium movement and accumulation in the affected environment.	• Provides central direction for the multiple initiatives related to modeling of cesium behavior in the environment.	• Identify key parameters and relevant to understanding impacts to human health and the environment. • Directly link model development to monitoring of the affected environment. • Calibrate models using observed data • Assess the selected models using monitoring and sampling data
2. Develop and apply models to evaluate and enhance the effectiveness of decontamination strategies and technologies.	• Provides a specific focus for modeling efforts that can lead to improved methods for decontamination methods.	• Evaluate mechanisms related to removal of cesium from urban environments. • Evaluate mechanisms related to remediation of agricultural lands including approaches to block biological uptake of cesium in agricultural products. • Evaluate mechanisms related to remediation of forest areas include a broad range of phytoremediation approaches.
3. Develop and apply models to inform urgent radiation protection strategies for people living in areas with residual contamination (re-entrainment) and for re-population of evacuated areas.	• Provides improved understanding of actual dose and risk contributors that should inform radiation protection strategies.	• Develop population dose/exposure models that are relevant to the specific circumstances, life-styles and exposure pathways that are actually present. • Evaluate potential recontamination mechanisms due to terrestrial transport of contaminants into areas that have undergone decontamination.

Table 1. (Continued)

Recommendations	Rationale/Benefit	Supporting Actions
Cesium Behavior in the Environment		
4. Develop and apply models to guide long-term monitoring approaches that will enhance the long-term understanding of cesium (and other contaminants) behavior in the environment.	• Ensures direct and frequent feedback between model development and monitoring results. • Helps guide monitoring system requirements by identifying likely pathways and accumulation points for cesium.	• Develop and maintain a direct interaction with the design of long-term monitoring efforts. Models should be used to define priorities for environmental sampling and analysis including identification of indicator species (sentinels), etc.
5. Investigate cesium effects on environmental receptors.	• Provides the basis for predicting the long-term human health and environmental impacts.	• Develop and maintain a dose - health dataset (that maintains individual privacy) to augment the international body of data used to evaluate health effects from radiation.
Remediation Strategy		
1. Conduct a systematic review of the decontamination work that has been completed to date (cost, effectiveness, waste generation, etc.) to provide the information base for extrapolating to implementation of remaining decontamination work.	• Provides a clear base of experience and information for informing national, prefectural and local decisions about remediation strategies. • Provides a basis for making realistic estimates of the remediation resources that are required and the most likely schedule requirements.	• Summarize the resources required to complete remediation efforts (cost, time, workers) and the impacts of the work (dose reduction, waste generation, etc.) and to be summarized on a city-by-city and prefectural basis. • Use this set of information to evaluate the relative effectiveness and efficiency of remediation efforts carried out by local communities and MOE-led areas to gain insights into the "best practices" that should be pursued in subsequent work.
2. Develop the baseline definition of the total set of decontamination work that needs to be completed.	• Provides the basis for tracking national progress toward completion of remediation efforts.	• Quickly develop a comprehensive approach for collecting and recording decontamination work progress, dose levels achieved, completion status, and waste generation and accumulation.

Remediation Strategy		
Recommendations	**Rationale/Benefit**	**Supporting Actions**
		There needs to be a national system to collect decontamination work results, progress, and remaining work. • Prepare a routine "progress report" that combines remediation results from all municipal, prefectural and national efforts. Should include cost to date, dose reduction achieved, decontamination waste generated, etc.
3. Develop and maintain an overall remediation strategy complete with life cycle cost estimates, resource allocation strategies (e.g., manpower, etc.), and analysis of critical strategic alternatives.	• Provides a resource-based strategy that reflects national, prefectural and local interests for completing remediation of the affected areas. • Supports a national dialogue regarding the best use of resources and the priorities that should be applied for remediation efforts.	• Conduct a "feasibility study" for defining and evaluating options for proceeding with full-scale decontamination in high dose areas. This should be conducted in parallel to the model demonstration projects (6 months of effort) that will be conducted in the high dose areas. • Use this strategy to guide resource allocation for the Intensive Survey Area (white zone) and the Special Decontamination Area (green, yellow and red zones). • Develop an end state definition for determining completion of remediation considering the potential diminishing returns and best use of remediation resources. This definition should be linked to the long term environmental monitoring information.
Cross-Cutting Considerations		
1. There is an immediate need to develop more effective processes for public involvement in remediation system decisions (e.g., site selection for treatment and storage facilities, re-population strategies for evacuated areas).	• Improvements in this area are necessary to support the very difficult challenges of restoring the evacuated areas.	• Public involvement practices and consensus building effectiveness have varied significantly across the communities undergoing decontamination. GOJ should review the variations in these practices and identify the factors that lead to success and that could be applied more broadly within the affected areas. • For the communities within the evacuation area (Special Decontamination Area) Consider instituting a community/ stakeholder involvement process (potentially formally chartered

Table 1. (Continued)

Remediation Strategy		
Recommendations	**Rationale/Benefit**	**Supporting Actions**
		working groups similar to U.S. citizen advisory boards at large cleanup sites) to provide advice on all aspects of remediation. Advice should support: • Community-specific input to radiation protection guidelines for re-population of evacuated areas • Priorities for implementing decontamination efforts • Waste management strategies including site selection for treatment and storage facilities • Other aspects of community reconstruction and restoration.

[1] ESF review and recommendations are focused primarily on handling of soil and waste generated as a result of decontamination work in Fukushima Prefecture. In general, “Specified Waste” in Fukushima and other prefectures is handled using separate treatment, storage and disposal systems. Specified waste includes tsunami debris and other disaster waste with cesium contamination greater than 8,000 Bq/Kg.

1.0. INTRODUCTION

Surface contamination of the lands from the accident at the Fukushima Daiichi Nuclear Power Plant (Daiichi NPP) site occurred in a population region and required evacuation of a population from the contamination regions. The predominant radioisotopes of the contamination, Cs-137/Ba-137m and Cs-134, were sufficient to cause a high radiation field in the air above the surfaces with a high effective dose rate that would require removal to enable a repopulation of the lands.

Early after the accident in November 2011, the Government of Japan (GOJ) Cabinet Office commissioned 11 Model Decontamination Projects that were led by joint ventures of contractors and coordinated by the Japan Atomic Energy Agency (JAEA). Those projects investigated technologies for decontamination of the surfaces of man-made structures and natural lands and decontamination methods specifically for the contaminated regions off-site from the Daiichi NPP reactor site. In parallel, the GOJ passed a law ("The Act on Special Measures") in 2011 that came into full effect on January 1, 2012 that effectively aligned its ministries and agencies to respond to the contamination for remediation of the lands. The Ministry of the Environment (MOE) was given the responsibility and authority for the off-site decontamination at that time. To date, much decontamination work has been performed and a look at the overall effectiveness of remediation activities has been performed.

By agreement with the GOJ MOE, the United States (U.S.) State Department, U.S. Department of Energy, and U.S. Environmental Protection Agency sponsored and deployed U.S. experts in radiological decontamination and environmental remediation on a mission to provide support requested by the MOE, including to provide recommendations for consideration in its work. The ESF mission period was February 4 through March 29, 2013. Pertinent biographical statements of the Embassy Science Fellow experts are contained in Appendix 1 to this report.

The radioactive material settled and caused contamination of a large portion of land spanning the Iwate, Miyagi, Fukushima, Ibaraki, Tochigi, Gunma, Saitama, and Chiba Prefectures. The vast majority of the contaminated land is in Fukushima Prefecture. The observations in this report are for the Fukushima prefecture only; the ESF team did not visit the other prefectures or review their specific decontamination plans.

The decontamination-related activities assessed by the ESF experts included the high-level organizational systems used for planning, execution, and regulation of remediation as well as the details of decontamination technology application and decontamination field practices.

Chapter 1 of this report summarizes the: 1) extent of initial contamination; 2) status of the decontamination activities; and 3) program elements of an environmental remediation system for a populated land with cesium contamination.

Chapters 2 through 8 describe the observations of the ESF team with respect to the environmental remediation program elements and offer both general and specific recommendations for improvements within the remediation program elements.

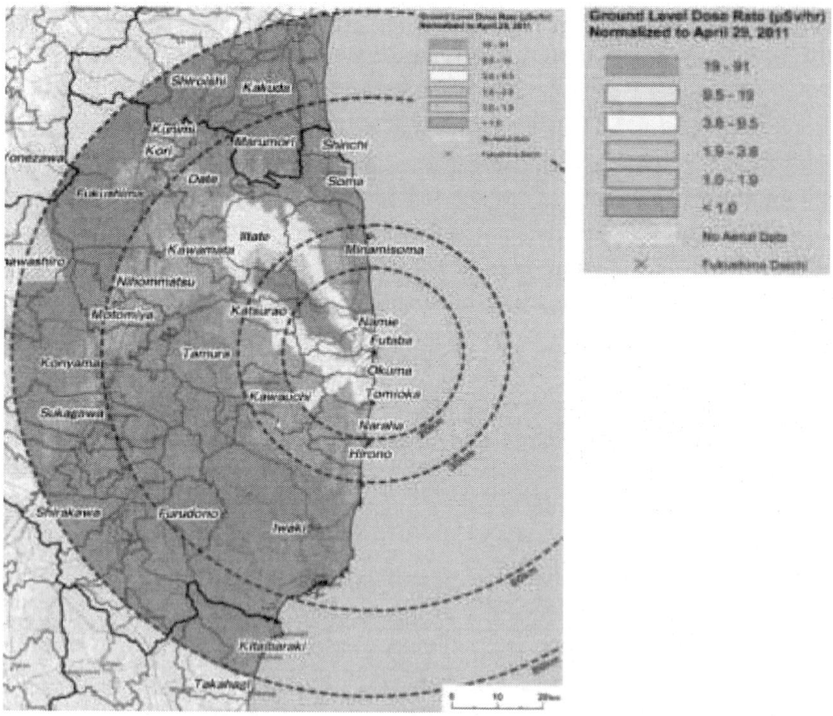

Figure 1-1. Aerial Survey Results – Joint US/Japan Survey Data, April 29, 2011.

1.1. Background-Contamination Caused by the Daiichi NPP Accident

The contamination released from the accident was surveyed aerially by a joint team from Ministry of Education, Culture, Sports, Science and Technology (MEXT), DOE National Nuclear Security Agency (NNSA), and US Joint Force (USJF) beginning on March 17, soon after the loss of containment on March 15. The data showed no measurable deposit of radiation after March 19, and the radiation levels since have been decreasing due to natural attenuation.[1] Figure 1-1 shows the composite results of the aerial surveys on April 29, 2011. A Restricted Area (exclusion zone) at a distance of 20 kilometers from the site was established, and a Planned Evacuation Area (special evacuation area) northwest of the site that corresponded to a general area with a dose rate of equal to and greater than 20 mSv/year was established. This combined area, called the Special Decontamination Area, is shown in red in the left portion of Figure 1-2.

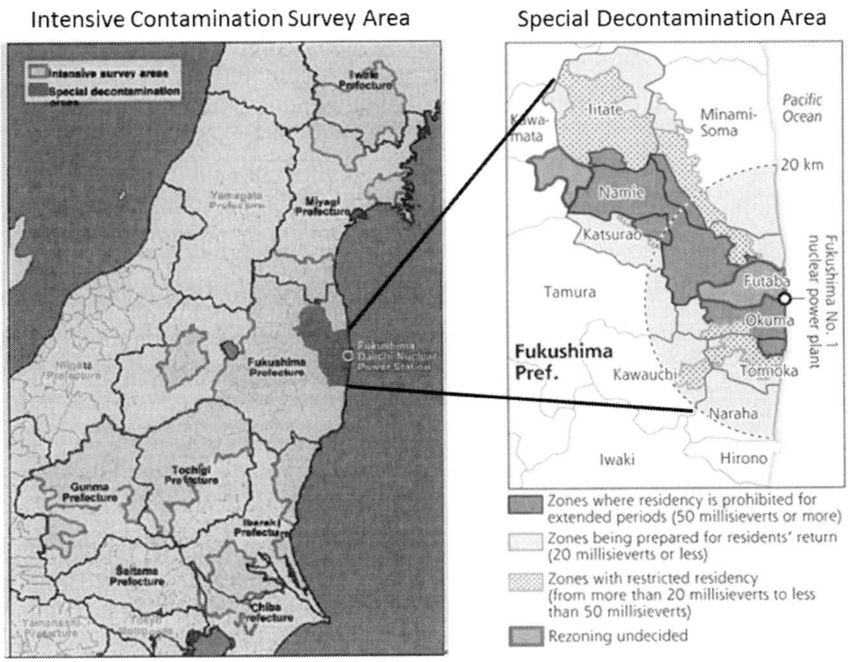

Figure 1-2. Map of the Special Decontamination Area and the Intensive Contamination Area in the Japan Prefectures.

The MEXT and JAEA continued to perform aerial and ground surveys, including those with low-flying aircraft. Correlations from measurements of radiation by the aircraft and to the ground contamination were established.

1.2. Background – Off-site Decontamination for Remediation

The Government of Japan-Cabinet Office commissioned the Japan Atomic Energy Agency to apply standard technologies (e.g., washing) for decontamination in 2 small test projects that began in August 2011 and ended in March 2012. Further pilot or model demonstration projects for decontamination technology began in November 2012. The projects for decontamination of 11 municipalities (cities, towns, villages) in 3 municipality groups each with a joint venture of contractors under a separate lead contractor (Taisei; Kajima; Obayashi) were led by the Japanese Atomic Energy Agency for the Cabinet Office of the Government of Japan. The purpose of the model projects, a clean-up of approximately 20 hectares (ha) per each municipality, was to establish a basis to demonstrate and evaluate decontamination.

The 11 pilot projects were completed in 2012. Pilot work included investigation and application and demonstration of decontamination technologies to surfaces and lands, dose monitoring, waste treatment for minimization, and waste storage (temporary storage).

Documentation of the decontamination pilot projects has been published in Japanese[2]. A recent JAEA website was has been set up to provide a description of the work done by JAEA[3]. Additionally, the JAEA led a total of 25 tests to evaluate novel or "advanced" decontamination technologies. These tests were completed in 2012.

With the assignment of the Ministries' roles under the Act on Special Measures, the Ministry of the Environment assumed responsibility under the GOJ for off-site decontamination work beginning in November 2011 and with full implementation on January 1, 2012[4]. The MOE organizational structure to deal with this new role includes field offices in the prefectures as shown on their web page[5].

The MOE developed a Decontamination Roadmap[6] that listed the timeline, beginning January 1, 2012, for the Decontamination Plan for the Special Decontamination Area. Figure 1-3 and Figure 1-4 below show the short- and long-term timelines, respectively, for the decontamination of the SDA.

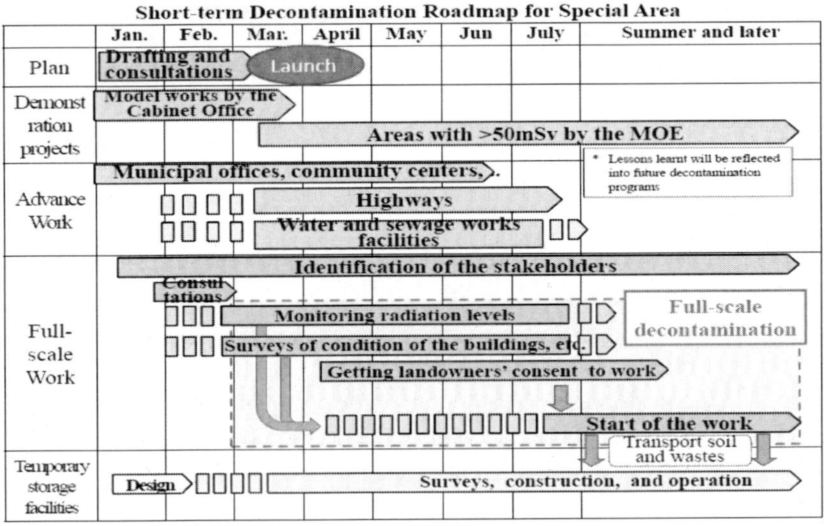

Figure 1-3. Short-Term Roadmap for the Decontamination in the Special Decontamination Area (for calendar year 2012).

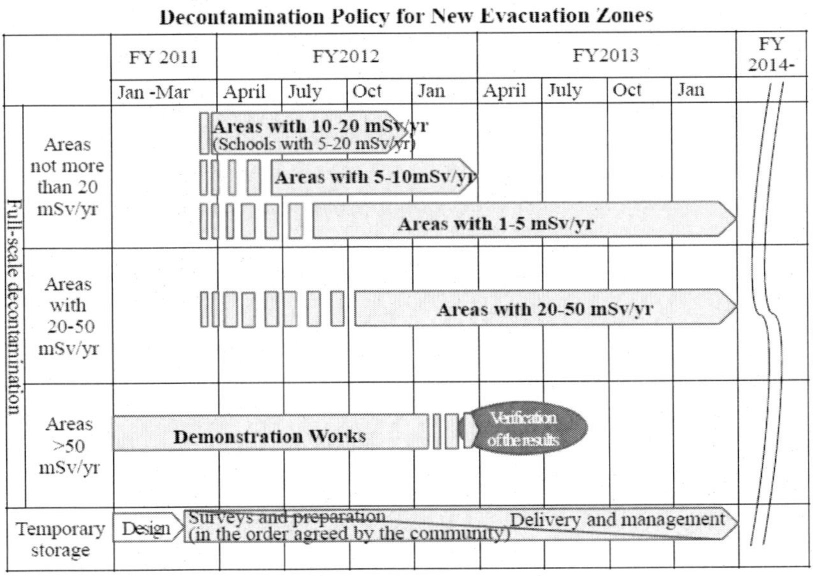

Figure 1-4. Long-Term Roadmap for the Decontamination in the Special Decontamination Area.

The SDA includes regions with contamination causing an effective dose rate of:

- Less than 20 mSv/year
- 20 to 50 mSv/year
- > 50 mSv/year.

The Roadmap also includes the schedule for the decontamination demonstration model project work, the early decontamination work (of municipal offices, community centers, the Joban Highway that is under the control of the MOE, and other key infrastructure for the municipalities), and the design and construction of Temporary Storage Facilities in the SDA that is concurrent with the initiation of the full-scale decontamination.

The Temporary Storage of the waste from decontamination work is planned for up to 3 years. The Interim Storage Facility (ISF), with a mission life of up to 30 years, would follow Temporary Storage. The ISF is scheduled to begin receiving waste in 2015 following site selection, procurement of land, and final design. ISFs in Futaba, Okuma, and Naraha towns are being planned to store a total expected volume of 28 million m^3 of waste. Permanent disposal of the wastes associated with the decontamination work is yet to be planned.

The roadmap also includes full-scale decontamination of the Intensive Contamination Survey Area (ICSA), which encompasses regions outside the Special Decontamination Area. The ICSA has contamination causing air dose rates less than 20 mSv/year. This work to decontaminate the ICSA is led by the municipalities with sponsorship and concurrence, and technical support as requested, from the MOE. For prefectures outside of the Fukushima Prefecture, the TSF and permanent disposal of wastes from decontamination would remain in their respective prefectures.

The MOE uses the following implementation strategy for decontamination in the SDA:

- The results from the model projects that demonstrated the technologies are used in the full-scale decontamination;
- Decontamination of municipal offices and school areas are the first decontamination work in a region;
- Agreements with the municipalities on Temporary Storage Facility locations and decontamination plans are secured before full-scale decontamination work begins.

The MOE has provided Decontamination Guidelines, Waste Guidelines, and Radiation Worker Guidelines based on the results and best practices of the decontamination work to date. These guidelines are the primary technical references for the decontamination work, and are provided to the municipal governments to aid them in the contracting and performance of their decontamination work as discussed below.

Overview and Status of the Remediation in the Intensive Contamination Survey Area

The Intensive Contamination Survey Area is the region outside the Special Decontamination Area where the contamination would cause an external effective dose rate of greater than 1 mSv/year and less than 20 mSv/year. The decontamination work is led by the local municipalities in conjunction with the prefectural governments. That is, municipalities develop the plans, report them to MOE for consultation, but municipality is the decision-maker. The GOJ-MOE provides funding and technical support, as requested, for the work. An overview of the overall decontamination progress in the Intensive Contamination Survey area of Japan, and its status as of August 12, 2012 is provided in reference 7.

Records are compiled and the status of the decontamination work in the municipalities is maintained by their prefectural governments. The decontamination status in the Fukushima Prefecture was provided to the ESF team by the government office in Fukushima City on March 7, 2013. Within Fukushima Prefecture, 40 municipal governments must conduct decontamination work in the Intensive Contamination Survey Area, and 36 of the 40 have developed plans and are implementing their plans. The status of number of homes, area of roadbeds, area of farmland, and area of forest (the 20 meters from houses) decontaminated and to-be decontaminated is available from the prefectural government.

The Fukushima Prefecture government also provides training to individuals performing the decontamination services (decontamination workers, field supervisors, and contract managers) in the prefecture.

Outreach activities are also provided through the Fukushima Prefecture government. A Decontamination Information Plaza is run by the Fukushima Prefecture and MOE. Models and demonstrations for radiation exposure and decontamination topics, radiation and decontamination information brochures, and technical expert staff are available to explain radiation topics to the public.

The Fukushima City municipality, approximately 60 kilometers away from the Daiichi NPP site, is one of the largest population areas in the

Fukushima Prefecture, and the largest amount of decontamination work, and volume of waste, has been generated to date in this area. The decontamination status for each of the districts in the city is reported separately.

There are 19 districts in Fukushima City. Dose rates in 2011 could yield an annual dose of 10 mSv/y or more in some districts, well above the GOJ's decontamination goal of 1 mSv/y. Priority in decontamination is given to "hotter" areas. For the work in Fukushima City districts to date, about 5000 homes have been decontaminated, and "at-location storage" of the waste is being implemented before Temporary Storage (Figure 1-5). About 4/5 of the homeowners have the decontamination waste buried on their home sites.

A dosimetry campaign using dosimeters for external radiation effective dose, and Whole Body Counting for the committed effective dose, was done for a segment of the Fukushima Prefecture population in Fukushima City, Nihonmatsu, Date, and Koriyama cities. The results of this campaign have shown that for most people, the actual doses received were well below the doses estimated by GOJ's standard dose calculation model (see Chapter 2).

Overview and Status of the Full-Scale Decontamination in the Special Decontamination Area

Full-scale decontamination of the Special Decontamination Area under the lead of MOE is in progress in four towns. The MOE web site contains up to date information of the status of the decontamination in the SDA.

Figure 1-5. In-Place Storage at a Private Residence in Fukushima City. This storage precedes the Temporary Storage that is planned for 3-year duration to be followed by Interim Storage that is planned for 30-year duration.

Figure 1-6. Shot-bead Cleaning of Sidewalk at the Sports Complex in Tomioka Town. This method is effective at removing near-surface with attached cesium contamination. This mature method is used in full-scale decontamination work.

Overview and Status of High-Dose Decontamination Demonstration

The organization of remediation areas within the SDA is shown in Figure 1-7 below, current as of end of March 2013.

Work to decontaminate the high dose rate regions (> 50 mSv/year) in the SDA has not begun. A special decontamination project in high dose rate region is being planned by the MOE. The project work would be performed in FY13. The decontamination of the surfaces using the common decontamination methods, and potentially advanced methods, will be used to evaluate decontamination effectiveness and waste generation. Future decisions on when and how to proceed with the full-scale decontamination of these high dose regions would be made by MOE following the completion of this project.

The Reconstruction Agency, a new GOJ agency expected to lead Japan in the reconstruction process by promoting and coordinating the policies and measures of the Government as well as supporting reconstruction projects to be implemented by the local municipalities, was established on February 10, 2012 from the "Act on Establishment of Reconstruction Agency." This agency will play a major role in the decisions to allow re- population in the SDA, and the decisions are understood to be based on ensuring that the infrastructure of the municipalities is safe and sound, and that the radiation dose rate conditions in the municipality are at the levels agreed to by the stakeholders, and the federal and local government bodies.

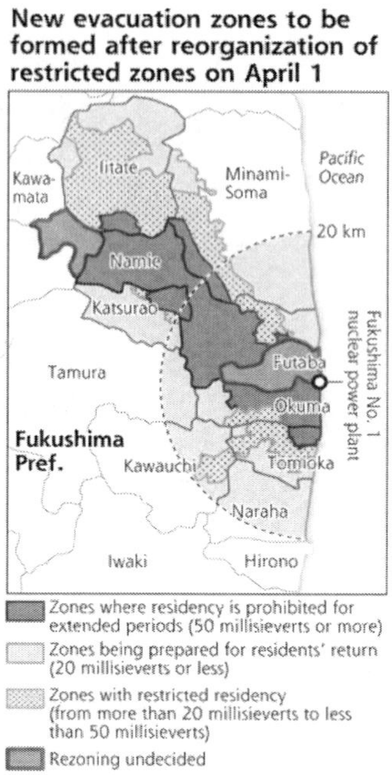

Figure 1-7. The Contamination Status of Land Regions and Evacuation Zone Planning by the GOJ-MOE as of April 1, 2013.

1.3. Program Elements for an Environmental Remediation System for a Populated Region Contaminated by Cesium

The information assessed by the ESF team included activities that the team expected to be in place for the remediation of wide-spread cesium surface contamination in a region including agricultural lands, urban areas, and forest regions. The ESF team formulated a framework (Figure 1-8) to organize the extensive information that was collected and reviewed and to formulate observations and recommendations.

The remediation of lands for re-population following radioactivity contamination is described in terms of high-tier elements that establish the framework for a total system description. In general, they can be applied to any radioisotopic contaminant. These high-tier elements are:

I. Radiation Protection
II. Decontamination Methods

III. Waste Management System
IV. Environmental Monitoring
V. Cesium Behavior in the Environment
VI. Remediation Strategy.

The Radiation Protection element involves the topics of selection of annual dose limits and dosimetry methods to demonstrate the dose to the public is within the selected limits. The Decontamination element involves the topics of decontamination technologies for the various materials/surfaces and their effectiveness. The Waste Management element involves the integrated activities of waste transportation, treatment, storage, and disposal that would be in place for waste management from its generation in the decontamination processes through its ultimate disposal. The Environmental Monitoring element involves the set of monitoring systems required for cesium presence in the environment including dose rate characterization until the time of its natural attenuation and effective removal from the environment. It also includes the effective reporting of the data and information. The Cesium Behavior in the Environment element comprises a body of investigations and studies to understand the migration of cesium both on made-man structures and in the biosphere, and evaluate it effects to biota either through uptake and committed exposure or to external radiation exposure. The Remediation Strategy involves the planning and strategy for the overall remediation work.

Figure 1-8 below shows the relationship of these elements necessary to promote the Remediation of the Environment Impacted by the Daiichi NPP Accident.

The protection of human health under the Radiation Protection element drives the decontamination targets and the work to achieve these targets under Decontamination. The Radiation Protection also drives the design of the Waste Management System to ensure radiological safety in their design such as control of contact dose at the boundary of the facilities, and robust design of the treatment, storage, and disposal systems such that premature failure and recontamination of the environment would not occur to cause significant exposure and impact on human health.

The Environmental Monitoring is needed to monitor for changing conditions such as for cesium accumulation points and for the reduction of radiation due to natural attenuation of cesium in the environment. Monitoring is essential to evaluate the present and trending dose rates that impact the Protection of Human Health since an estimation of annual dose received by an individual is made by the results of a monitoring activity. Further monitoring of cesium at natural accumulation points in a region can detect if the accumulation is occurring and further, at sufficient accumulation, a need for further Decontamination.

Input into Radiation Protection is the information on cesium effects on human health that would include any new information from an international community, and the investigations that will be occurring under the investigations performed in the Cesium Behavior in the Environment element.

Overall Remediation of the Environment effected by the Daiichi NPP Accident is defined by the new (post-Decontamination) conditions, an understanding of Cesium Behavior in the Environment, and Environmental Monitoring throughout the time until cesium has been sufficiently attenuated from the environment to no longer pose a health risk or potential to adversely impact the environment.

The observations and comments in this chapter are on the remediation work done to date, and the work that is planned to date. The information was gathered from meetings and field observations made by the ESF team of part of the observations.

The plans, technical reports, and status reports that were provided by MOE and the other agencies and reviewed as part of this work are listed in Appendix 2.

The ESFs also provided information from their respective U.S. sites that includes U.S. information, lessons learned, and potential technologies that may benefit the remediation activities in the off-site decontamination. A listing of the information provided to the MOE from the U.S. is contained in Appendix 3 to this report.

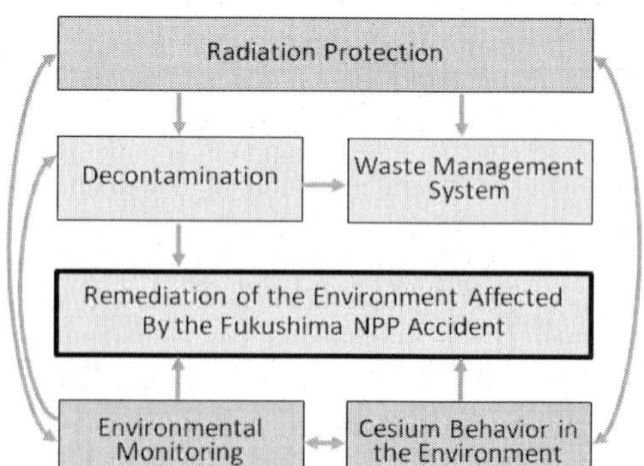

Figure 1-8. Program Elements for an Environmental Remediation System for a Populated Region Contaminated by Cesium.

The following chapters correspond to the remediation system elements. The subsections are organized by:

- Current Situation
- Summary of Observations
- Recommendations and Supporting Actions.

A comprehensive itemized listing of the recommendations, rationales/ benefits, and supporting actions for each element of a total system remediation project is provided in Table 1.

2.0. Assessment of Remediation System Program Elements: Radiation Protection

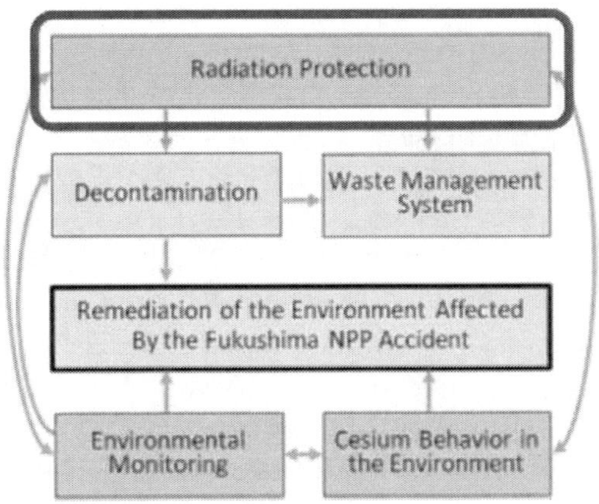

2.1. Current Situation

Radiation Protection is the portion of the remediation process that focuses on protection of human health. Radiation protection helps to: develop practices aimed at protecting people who are living in contaminated regions; set goals or targets for decontamination activities; and develop guidelines for re-population (or reoccupancy) of currently evacuated areas. Re-population decisions must balance a complex array of factors that define the social and economic benefits of returning to previously evacuated areas. To support re- population decisions for Fukushima NPP incident there is a definite need to refine the wide gap between the long-term cleanup goal of 1 mSv/y and the evacuation standard of 20 mSv/y. MOE has been developing decontamination guidelines and implementing them in various regions to protect human health. The government of Japan is planning to establish radiation protection measures currently under consideration for the evacuees' return by the end of this year.

Policy for Decontamination

The contaminated areas are broken down into two categories under the "Act on Special Measures": the Special Decontamination Area and the Intensive Contamination Survey Area. The government of Japan set the long-term cleanup goal for both areas. A 1 mSv/year level is the standard used by Japan for protection of the public from more typical radiation exposures.

Special Decontamination Area (SDA, Restricted Area and Deliberate Evacuation Area): The government of Japan ordered residents to evacuate the area within 20 km of the NPP site (Restricted Area) on March 12, 2011 immediately after the first explosion at the Daiichi plant. The Deliberate Evacuation Areas were designated as the areas where the annual cumulative dose was expected to be more than 20 mSv on April 2, 2011. Two evacuated areas are shown in the Figure 2-1 as Restricted Areas and Deliberate Evacuated Areas in red. Approximately 86,000 residents in 14 municipalities were affected by this evacuation.[8] The evacuated areas are currently separated into three different areas as shown in Figure 2-2. These areas are defined according to their annual cumulative doses: Area 1 (<20mSv/y), Area 2 (20-50 mSv/y), and Area 3 (>50 mSv/y). The national government (Ministry of the Environment or MOE) is responsible for remediation in SDA. Full scale decontamination in SDA will be carried out by MOE, whose goal is to reduce the additional exposure dose to 1 mSv/year as a long-term goal. The decontamination goal for Area 3 will be determined based on the demonstration project results (scheduled later in FY 2013). As of end of March 2013, 9 of the 11 Municipalities (4 of them with decontamination in progress) in SDA have prepared their decontamination plans except the town of Tomioka (local coordination in process) and Futaba town.[9] The detailed information on the remediation progress is available in the "Remediation Strategy" section of this report.

Intensive Contamination Survey Area (ICSA): Outside of the Special Decontamination Area, 101 municipalities in 8 prefectures (designated as the Intensive Contamination Survey Area) exceeded the annual cumulative dose of 1 mSv/year. In the ICSA, decontamination efforts are in progress by individual municipalities with the technical and financial support from the national government.[10] Most (92) municipalities in this area have developed statutory decontamination plans as of January 7th 2013. The communities in ICSA are carrying out their daily activities while decontamination work is underway. Figure 2-3 shows the ICSA outside the SDA.

Standard Equation for Estimating Annual Dose Rate

The government of Japan uses a standard calculation model to estimate annual exposure from hourly ambient dose rate measurements for the public. This standard model is based on the assumption that people spend 8 hours a day outdoors and 16 hours a day indoors. The dose reduction due to shielding effects of being indoors is assumed to be 40%. This standard model with current assumptions tends to lead to higher exposure estimates than actual individual exposure. This observation is based on the results from the individual dosimetry data shown in Figure 2-5. MOE's remediation goals are also assessed using this standard calculation model.

Figure 2-1. Special Decontamination Area.

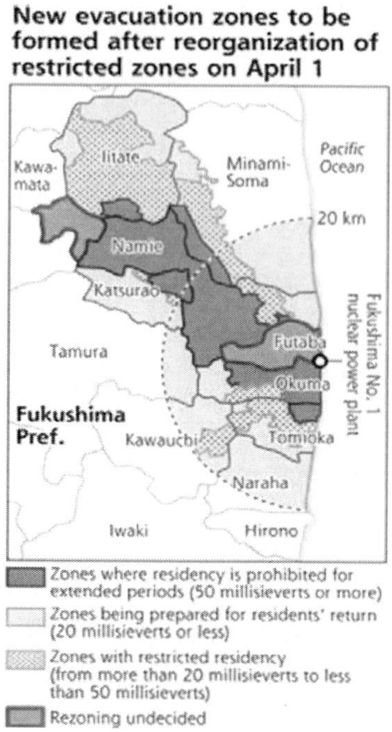

Figure 2-2. Current status of the evacuated areas.

A dose rate estimation model for a representative person has been developed and has been published by MOE in a brochure for the public[11]. This model converts the ambient dose rate in μSv/hour to an annual dose that is used to compare against the international guidance for the yearly dose. Figure 2-4 below describes how the model is applied.

This model is an estimation of the annual dose that would be received by a "reference" individual living in an area with a given measured airborne dose rate. This is used as a general guide for planning radiation protection and decontamination activities. It is not accurate for any specific individual because dose rates are not uniform across the area, people spend varying amounts of time indoors, some people consume food from their own gardens, and shielding is not uniform in all buildings, etc.

Prior and Existing Radiation Protection Efforts

Fukushima Health Management Survey: The GOJ has been conducting the radiation exposure assessments, through the Fukushima Health Management (FHM) Survey[12], for the residents in Fukushima Prefecture. The

FHM survey is conducted by Fukushima Medical University with the support of Fukushima Prefecture for the people who were living or present in Fukushima Prefecture on March 11, 2011. The FHM survey has two components, the basic survey and the detailed surveys. The basic survey targeted the entire population in Fukushima Prefecture to estimate individual radiation exposure levels in the 4 month period from the nuclear accident at Fukushima Nuclear Power Plant (March 11th, 2011 until July 11th, 2011).[13] The detailed surveys are composed of the following programs: thyroid ultrasound examination targeting residents aged 0 – 18 years as of March 11th, 2011 (conducted May 2012 – March 2013), comprehensive health checkup for former residents of evacuation zones (conducted April 2011 – March 2012), mental health and lifestyle surveys for former residents of evacuation zones, and pregnancy and birth survey for women who received Maternal and Child Health Handbooks from municipal offices in Fukushima Prefecture between 1 August 2010 and 31 July 2011, and those who had handbooks issued in other prefectures but received prenatal care or delivered babies in Fukushima Prefecture after the disaster.[14]

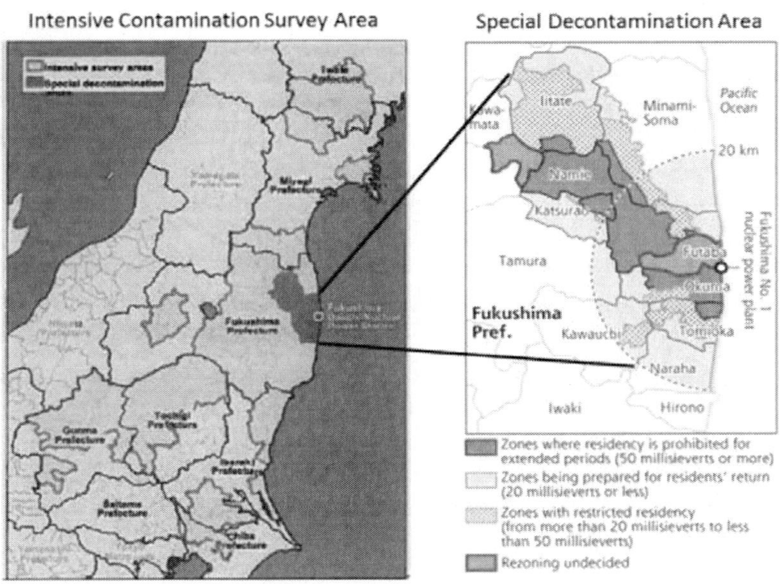

Figure 2-3. Intensive Contamination Survey Area in yellow and evacuated area.

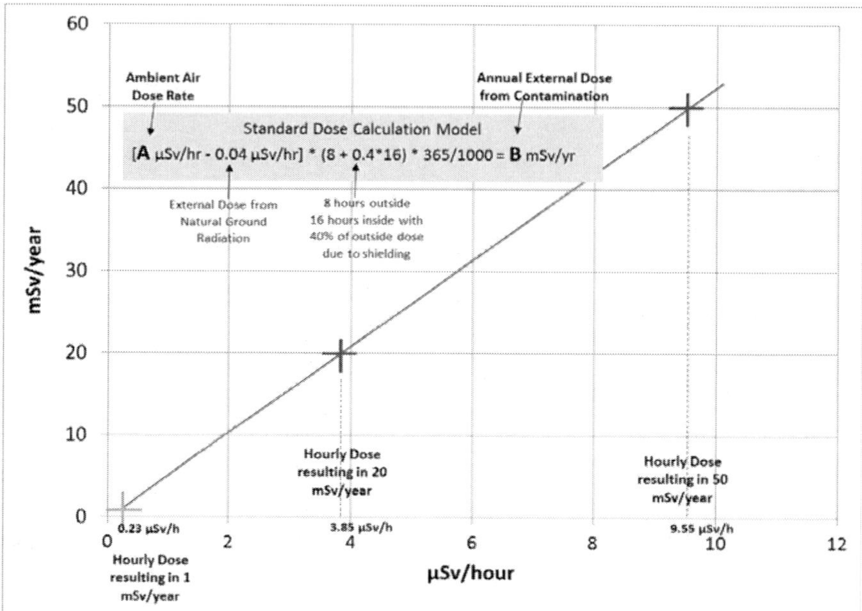

Figure 2-4. Standard calculation model to estimate annual exposure from hourly ambient dose rate measurements.

Using the basic survey results, the effective external cumulative doses were estimated for the survey respondents (approximately 400,000 as of January 31st 2013). The survey results showed that the estimated external cumulative doses were less than 5 mSv for 99.8% of the respondents (excluding radiation workers).[15] Approximately 150,000 children under age 18 have participated in the thyroid ultrasound examination as of January 25th, 2013. The examination was conducted by identifying the size of nodule and cyst at two different periods (April, 2011 – March, 2012 and April, 2012 – January, 2013). Of these, 735 participants required the secondary examination because their nodule sizes were equal or larger than 5.1 mm or cysts were equal or larger than 20.1 mm.[14] Approximately 92,000 people participated in the mental health and lifestyle survey and about 5 % of the people responded that they required support. Out of 16,000 questionnaires, more than 9,000 women responded to the pregnancy and birth survey, and 1,393 respondents were counseled by midwives and public health nurses through August 31st 2012.[14]

Whole-body Counting: The internal radiation doses have been estimated for the residents living in the potentially high dose areas using whole body counters. This program was organized by Fukushima Prefecture, Niigata Prefecture, and Aomori Prefecture in collaboration with National Institute of Radiological Sciences, Japan Atomic Energy Agency, and Minami Soma city hospital. More than 90,000 residents were examined through October 31st, 2012, The results showed 99.99% of participants had received less than 1 mSv of internal cumulative dose (with two residents receiving 3 to 4 mSv).[16]

Children Thyroid Exposure Screening Survey: The Nuclear Safety Commission (NSC) of Japan conducted preliminary estimation of thyroid equivalent dose for children using the System for Prediction of Environmental Emergency Dose Information (SPEEDI)[17] and monitoring data on March 23rd, 2011. The results showed some areas (northwest and south of NPP) had relatively high thyroid doses compared to the evacuated area. Requested by NSC, the local Nuclear Emergency Response Headquarters conducted monitoring for thyroid dose level for 1,149 children in Iwaki city, Kawamata town, and Iitate village from March 24th to March 30th 2011. The screening survey was conducted using NaI scintillation survey meters. The results showed that the thyroid dose rate was less than the screening level (0.2 μSv/hr, corresponding to a thyroid equivalent dose of 100 mSv) for all examined children (for the age group of 1 year old as of March 24 2011) and the maximum dose measured was 43 mSv.[18]

Dosimetry Program: Dosimetry program information was provided by the Fukushima city government during the meeting on March 7th 2013. The dosimeters were prepared for the pregnant women and children (0-15 years old), a total of 48,700 (approximately 37,000 children). The dosimeter measurements were taken for two different periods: September 1st – 30th 2011 and October 1st –November 30th 2011. The data were reviewed by the Fukushima City Health Care Review Board, which is composed of doctors from municipal medical association and municipal radioactivity advisors. The Review Board concluded that there is low probability of future cancer risk due to radiation. Comparison of measured and estimated external cumulative radiation is shown in Figure 2-5.[16] The program was conducted during two different periods: September – December of 2011 and December 2011 – February 2012. Radiation dose was measured for residents in four cities (Nihonmatsu, Fukushima, Date, and Koriyama) using dosimeters. The number of participants was 8,725, 36,767, 8,505, and 25,551 from Nihonmatsu,

Fukushima, Date, and Koriyama, respectively. For all participants, cumulative doses, as measured by the dosimeters, were less than the estimated doses based on the standard model used. Currently, a dosimetry program called "Residents Monitoring Service by Light-ct56" is available for former residents of evacuation zones who move back to their hometown.

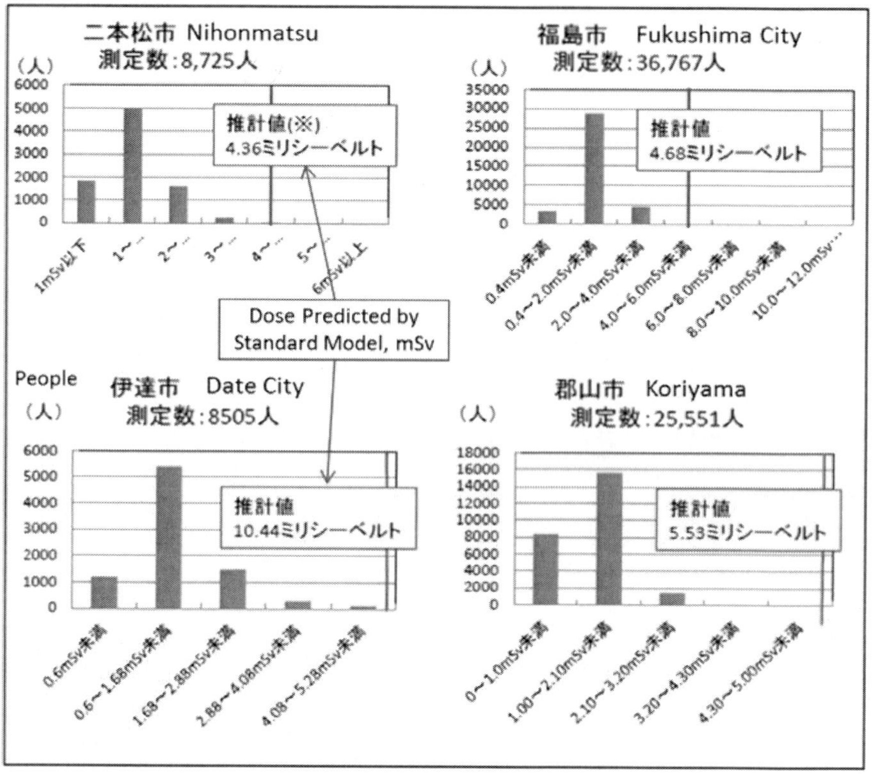

Distribution of Measured Doses (from dosimetry data) Received by Residents in Four Cities Compared to Predicted Dose Based on Standard Model (Source: http://www.reconstruction.go.jp/topics/post_132.html)

Figure 2-5. Distribution of measured doses (from dosimetry data) received by residents in four cities compared to predicted dose based on standard model16.

2.2. Summary of Observations

The decontamination policy established by MOE to achieve the long-term goal for an effective dose rate of 1 mSv/year or less, for the SDA and ICSA is consistent with International Commission on Radiological Protection (ICRP) guidance provided in their 2009 report as a typical value used in other situations.[19] Since it may take several years to meet the long-term cleanup goal in some regions, there is a growing need to establish timely re- population and recovery approaches for the evacuated areas in a scientifically defensible manner. The residents in SDA have been evacuated for more than 2 years, and as a result there is a growing demand from the local governments about how to advise the citizens on re-population decisions.

The current GOJ standard dose calculation model used is considered to be too general to apply to specific individuals and lacks the flexibility needed to accommodate the circumstances for various population groups. The standard calculation model for estimating annual dose from an hourly dose rate makes assumptions that are not generally applicable to all population groups and local situations. This model may underestimate exposures for those who travel to more contaminated areas outside the home, eat contaminated foods, drink contaminated water, etc. This approach is reasonable for basic epidemiologic analyses, but a more refined approach can be developed using a combination of dosimetry, questionnaires, and studies. A more flexible model or approach would improve radiation protection planning for re-population of the SDA. The annual dose values estimated using the method of air dose rates at 1 meter height in residential and public areas are useful as an indicator to determine the community-wide decontamination progress. However, the individuals who have moved into these areas may be exposed to different levels of radiation due to the presence of hotspots and different lifestyles.

Various programs have been conducted to assess the radiation exposure of the population in the impacted area. These programs have focused mainly on the early and intermediate phase of the incident and its impacts on population groups. These programs are currently on-going project to monitor the potentially affected population and provide the necessary support to them. ESFs have not been informed about the detailed exposure assessment procedures for the whole body counting but in general, the whole body counting results must be used with caution due to the high probability of missed doses to the time since exposures, unknown counting time, and the effective half-life. The dosimetry program was conducted for children and pregnant women in 2011 and 2012. ESFs were not informed on how the dosimetry program will be implemented for the returning evacuees to the SDA.

The national government has developed various information booklets to inform the local governments, stakeholders, and general public on how to reduce further radiation dose under the normal living routines. The Decontamination Information Plaza in Fukushima City provides both a central facility and an outreach service for information regarding radiation, its effects on human health, and decontamination practices. The Plaza also provides the status of decontamination work in the Fukushima Prefecture. However, ESFs did not observe local communities and stakeholders to provide their inputs into the re-population decision making process.

2.3. Recommendations and Supporting Actions

Recommendation 1

Develop re-population and dose reduction framework and implementing process for application at a community specific level.

It is recommended that the GOJ develop a set of re-population guidelines and dose-reduction guidelines. The re-population guidelines should establish the conditions under which population of a municipality can return to their residences. The companion dose-reduction guidelines should then provide a plan that aims at dose exposure reduction over time once the area is re-occupied.

Supporting Actions

- Establish an Expert Advisory Group on radiation protection to support and interact with prefectural and municipal government officials, and to provide necessary information to the public and stakeholder groups.
- Develop guidance for remediation end-states (e.g., "how clean is clean?" or remediation completion criteria) and re-population standards and approaches that recognizes the unique characteristics of the incident and local considerations. Consider use of a phased approach for re-population efforts where the critical service providing personnel who return to the evacuated areas (e.g., public service providers, firemen, police, etc.) wear dosimeters to validate the actual dose exposure conditions in the community.

- Develop a dose estimation process that accounts for specific land uses and population groups (e.g., industrial, residential, schools, etc.). For example, using a tool such as DOE's RESidual RADioactivity (RESRAD) or EPA's Preliminary Remediation Goals (PRG) or Dose Compliance Concentrations (DCC) along with actual dosimetry data would provide a more useful estimate of potential dose to population groups.
- Include a community/stakeholder involvement process (formally chartered working group) to ensure that local considerations are incorporated into radiation protection processes.
- Identify institutional controls (e.g., avoidance of "hot spots," restrictions on access to areas that have not been decontaminated, etc.) that support ALARA principles for managing individual dose rates.

Guidance for the decision-making process that can be used to create dose rate re-population and dose-rate reduction guidelines are available from the USG. For example, the US EPA has developed the draft Protective Action Guides (PAG) Manual to provide guidance on protective actions and when to take them for early, intermediate and late phases of radiological incidents. The relevant resources are available in Appendix 3.

The ESFs also recommend using ICRP Publication 111 (Application of the Commission's Recommendation to the Protection of People Living in Long-Term Contaminated Areas after a nuclear Accident or a Radiation Emergency) as an important resource for developing re-population guidelines and associated radiation protection strategies. This report clearly recognizes that the re-population decision is not solely a matter of radiation protection, but must also balance all relevant factors including economic, social, cultural considerations, and more. The need for public transparency and the role of stakeholder involvement is clearly recognized and discussed.

Recommendation 2

Establish a radiation dosimetry program for residents who return to evacuated areas to provide the best information possible for understanding and managing population radiation exposure.

A dosimetry program should be used to directly quantify individual exposure in re-population areas. Dosimetry information can be used to develop improved dose reduction practices. The standard dose calculation model for estimating the annual dose that a reference individual would receive is useful for planning purposes. However, the standard calculation model has several limitations. A more representative model that can better differentiate specific population groups (school children, factory worker, farmer, etc.) should be deployed to provide more accurate information for decontamination planning and radiation protection practices for re-population. Additionally, the recommended dosimetry program would help determine actual radiation exposure for the various population groups and would support effective dose reduction practices for the population.

Supporting Actions

- Provide Expert Advisory Group and community/stakeholder interest groups to design and review the dosimetry program for returning population in the evacuation required areas. These groups should work collectively with government decision makers (national, prefectural, and community) to provide advice and to work collectively to resolve differences.
- Implement dosimetry program and regularly collect data for review (Recommendation #3).
- Conduct a systematic review of existing decontamination, health survey data, monitoring information including dosimetry data from affected population groups in Fukushima Prefecture. (There are data collected from the previous dosimetry program in multiple cities in Fukushima Prefecture in 2011 and 2012.)

It may be useful to continue and expand the external dosimetry program and internal whole body counter programs to ensure that the population that may be expected to receive doses are given the opportunity to obtain dosimeters and dosimetry records of their exposures. It is also suggested that the dosimetry program includes the following aspects: (1) a segment of the community to better characterize the public exposure, (2) random temporary environmental monitoring stations, (3) targeted environmental monitoring activities (i.e. local groceries, hotels, homes, community centers, entertainment/shopping centers, etc.), and (4) experts in exposure/dose assessment and epidemiology. The relevant resources can be found in the reference documents.[20, 21] The national and local governments can collaborate to proceed with this program in consultation with the Expert Group outlined in recommendation #4. This collaboration of governments and Expert Groups is important because the Expert Group inputs will improve the public trust and provide more accurate data that support decisions and response.

With periodic review of the outputs from the recommendation #3, the information from the dosimetry program can be used to verify and revise the guidance and further the dose estimation approach used. The periodic revision of this guidance is recommended using the inputs from automated dose rate monitors and dosimetry program as the remediation goes forward. Radiation workers at commercial NPP sites, and at DOE sites in the US are subject to an external and internal dosimetry program, and further information and examples for creating an efficient program for large (> 1,000) population groups are available based on this type of program and can be made available.[22,23]

Although collection of external and internal dosimetry information can be costly, application of improved dose estimation models would allow program cost reductions and can provide an accurate and reliable estimator for dose. Confidence in dose estimation would come from field measurements and comparison of the dosimetry results to dose estimation model results. There are various tools available from the international community for dose estimation. These include RESidual RADioactivity (RESRAD) developed by the US DOE (Argonne National Laboratory), Radiation Preliminary Remediation Goal (PRG) Risk Calculator and Applicable or Relevant and Appropriate Requirements (ARAR) Dose Compliance Concentration (DCC) Calculator used in Comprehensive Environmental Response, Compensation, and Liability Act (CERCLA) process by the US EPA. These tools each contain default input parameters for multiple land uses and allow users to adjust the default input which would allow using assumptions that are more appropriate for Japan. Detailed information regarding CERCLA process and related trainings are available in Appendix 3. Population groups can be constructed to reduce the variability from one large population group in dose received. For example, the population groups can be constructed as different age groups and further into types of profession. The dose rate for the specific group in the location can be predicted by using the dose estimation tools with the inputs from the monitoring data in the target area. The radiation reduction guidance can be developed based on the dose prediction. It is also recommended that this guidance consider the potential economic and social benefits of re-population depending on the local characteristics and situations.

Recommendation 3

Regularly review environmental monitoring results, dosimetry results and impacts from decontamination efforts to adapt the framework in Recommendation #1.

Timely update and review of the data collected from various measurements will ensure that the re-population and dose-reduction guidelines (from Recommendation #1) are effective and adjusted based on actual conditions and experience. As experience and data are gained, revisiting re-population process and guidelines may be required to be adjusted and optimized.

Supporting Actions

- The repopulation and dose reduction guidelines should be revisited as progress in remediation is achieved and as progress in community restoration occurs to ensure decisions made are effective.
- The review process should include the Expert Advisory Group (Recommendation #4) and community/stakeholder interest groups.

An example of community/stakeholder interest groups is Community Advisory Group (CAG) in US. The CAG is made up of members of the community and is designed to serve as the focal point for the exchange of information among the local residents and EPA, the State regulatory agency, and other pertinent government agencies involved in cleanup of contaminated site. The objectives, functions, membership and scope of CAGs are addressed in the US EPA guidance, Guidance for Community Advisory Groups at Superfund Sites (OSWER Directive 9230.0-28). Further resources are available in the following link: http://www.epa.gov/superfund/community/cag/resource.htm. While ICRP publication 111 provides general guidance on this topic, this US EPA example can provide more specific directives related to this topic. The GOJ can use this type of guidance as reference to develop the Japan specific community/stakeholder interest groups to improve the involvement and communication between governments and stakeholder groups. The USG has numerous experiences in promoting stakeholder involvement during the remediation processes, not only during radiological remediation, but also other remediation associated with disasters such as the 9-11 World Trade Center attack, Hurricane Katrina, the recent Hurricane Sandy. Depending upon GOJ interests, USG can provide relevant resources.

Recommendation 4

Establish Expert Advisory Group on radiation protection to provide technical assistance to prefectural and municipal government officials, and to provide necessary information to the public and stakeholder groups.

It is recommended that an Expert Advisory Group be established to provide the technical bases to the community/stakeholder interest groups. So the community/stakeholder interest groups can effectively provide their inputs for decision making process for re-population, and the guidelines for continued reduction in dose rates with time.

Supporting Actions

- The independent Expert Advisory Group should include selected National, Local, and private sector subject matter experts in fields as health physics, radiation protection, cost and risk analysis, remediation protection, and relevant regulatory requirements. This group periodically provides multi-disciplinary expert input on the radiation protection guidelines including technical issues, analysis of data, analysis of regulatory requirements, cost analyses and risk analyses (Recommendations #1 - #3)

The example of Expert Advisory Group is the "Technical Working Group" in the draft EPA's Protective Action Guide (PAG) Manual (http://www.epa.gov/radiation/docs/er/pag-manual-interim-public-comment-4-2- 2013.pdf). The draft PAG described the Technical Working Group as, "A technical working group should be convened as soon as practicable, ideally within days or weeks of the incident. The technical working group would be managed by the Planning Section Unit that is assigned responsibility for the cleanup options analysis." The draft PAG manual described the functions; multi-disciplinary expert inputs on the cleanup options analysis, including advice on technical issues, analysis of relevant regulatory requirements and guidelines, risk analyses and development of cleanup options. The manual specifies including the technical working group experts in the following areas: environmental fate and transport modeling, risk analysis, technical remediation options analysis, cost, risk and benefit analysis, health physics and radiation protection, construction remediation practices and relevant regulatory requirements.

Another example is Environmental Management Site-Specific Advisory Boards (EMSSABs) funded by DOE. The scope of activities of the EMSSAB is defined as "At the request of the Assistant Secretary or the Field Managers, the Board may provide advice and recommendations concerning the following EM site-specific issues: clean-up standards and environmental restoration; waste management and disposition; stabilization and disposition of non-stockpile nuclear materials; excess facilities; future land use and long-term stewardship; risk assessment and management; and clean-up science and technology activities." As an example, at the Savannah River Site, at the request of the DOE Assistant Secretary of Environmental Management or the SRS Manager, the Board may provide advice and recommendations concerning the following Environment Management site-specific issues:

- cleanup standards and environmental restoration;
- waste management and disposition;
- stabilization and disposition of non-stockpile nuclear materials;
- excess facilities;
- future land use and long term stewardship;
- risk assessment and management; and
- cleanup science and technology activities.

The EMSSAB may also ask, subject to Environmental Management approval, or be asked by Environmental Management, to provide advice and recommendations on any other Environmental Management project or issue. For the EPA's Superfund program, the cornerstone of EPA's efforts to provide technical assistance to communities is the Technical Assistance Grant (TAG) program. Under the TAG program, community organizations can apply for an initial grant of up to $50,000. TAG recipients use these funds to hire their own independent technical advisors to assist them in reviewing site related documents, and developing comments on a variety of areas such as sampling plans, risk assessments, proposed cleanup plans, ongoing operation and maintenance activities, etc.

A corollary to the TAG program is the Technical Outreach Services for Communities (TOSC) program. The TOSC program was started in 1993, and through it the EPA's Superfund program funds consortiums of universities around the country to provide independent technical assistance to communities. Communities also need procedural ways to be involved. In 1995, EPA started the Community Advisory Group (CAG) program. Through the CAG program, EPA seeks to bring together early in the remediation process a broad group of stakeholders (e.g., residents, local business owners, local government officials and others), who are interested in contributing at the EPA Superfund site.

3.0. ASSESSMENT OF REMEDIATION SYSTEM PROGRAM ELEMENTS: DECONTAMINATION

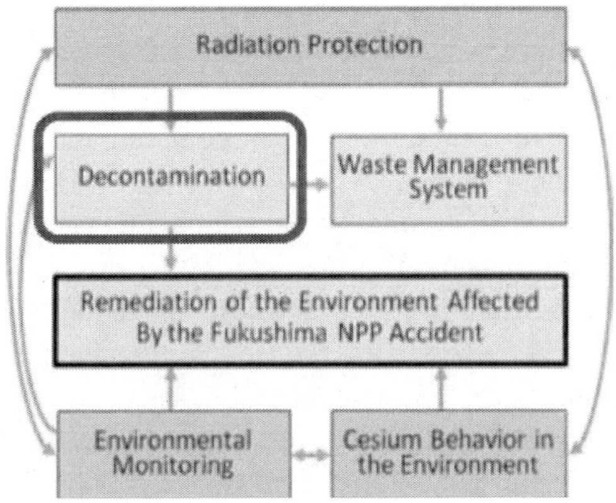

The radiological contamination of a solid material, in general, can be volumetric or surface, have multiple radioisotopes, and have very low to very high concentrations of the contaminants. The contamination can be physically bound and/or chemically bound to the material. The contamination may be selectively partitioned to a constituent or to a phase in the host material.

3.1. Current Situation

The radioactive cesium that has settled on the infrastructure and the natural regions in the prefectures near the Fukushima site has caused surface contamination. The contaminated areas and materials include:

- Roads of various material types and designs
- Soils for agriculture, playgrounds, other
- Grassy fields (meadows)
- Home lawns
- Home rock landscape
- Building structures of various material types and designs
- Roofs of various material types and designs
- Forests.

The level of cesium contamination is sufficient to produce high air dose rates at and above the surfaces. The primary aim of the decontamination work is to reduce the air dose rate at and above the surfaces to reduce the effective external dose rate to an individual at that location.

Ideal decontamination would be 100% or full removal of the contaminant from the material. In the case of surface contamination, total removal of the surface, down to the maximum depth of contaminant penetration, would provide full decontamination. However, it may not always be feasible, or even necessary, to remove the full depth of a contaminated surface of a structure, or the full depth of contaminated soil, to reduce the air dose rate. For example, a house that is highly contaminated and where the contamination is trapped several millimeters into the surfaces may not be aesthetically acceptable or even structurally sound after removal of the surface layer. Also, removal of soil to a depth to ensure 100% removal of the cesium may lead to excessively large volumes of waste that would not be feasible to manage.

There are drivers to investigate advanced decontamination technologies. Advanced decontamination technologies would lead to applied decontamination methods to extract the contamination from a surface and:

- leave the surface essentially intact;
- remove more of the contamination from a surface;
- achieve decontamination more cost-effectively; and/or
- achieve decontamination more quickly than the present methods.

The handling, transportation, storage, and ultimate disposal of the total waste produced from the decontamination should be considered in the evaluation of the cost-effectiveness of a decontamination technology with an applied decontamination method.

Several decontamination technologies with applied decontamination methods have been used in the decontamination work to date. A variety of advanced decontamination technologies have been investigated to date.

3.2. Summary of Observations

Field decontamination work has been performed in pilot (model) decontamination projects[24] and in full-scale decontamination work in the ICS area. Advanced decontamination technology investigations have been completed in a set of 25 decontamination technology demonstration test projects led by JAEA[25], and in a set of 22 decontamination technique demonstration projects led by MOE[26]. Additionally, 15 other advanced decontamination technology demonstration projects led by MOE are in progress. The decontamination work for both the pilot, full-scale, and advanced decontamination technology development are performed by Japanese corporations or joint ventures following process of proposal solicitation and selection and award of the work by MOE.

The GOJ-MOE has prepared Decontamination Guidelines[27] for the decontamination methods that they have evaluated and approved for decontamination work. Part 2 of the Decontamination Guidelines includes a list of decontamination methods for: buildings (roofs, gutters, wall, gardens, fences , etc.) that are basically debris removal and cleaning; roads (roadsides, drains, paved surfaces, unpaved roads) that are basically debris removal and cleaning; soil (schoolyards, farmlands) that are basically either tillage or soil removal; vegetation (lawns, trees, forests) that are basically removing the litter. These guidelines do not contain information on wipe media, water per unit area and time, but do contain some other information such as typical water pressure of 15 MPa, and soil depth removal in endnote documents. That is, this is the key document MOE gives to municipalities for ICS area clean-up and is available for their SDA clean-up.

The Decontamination Guidelines are given to the municipalities in the ICS area to provide the allowable methods in their contracted decontamination work. Although this is the key document that MOE gives to the municipalities, the municipalities may contract other methods with approval by MOE for sponsorship. Additional specifications on decontamination that the MOE uses to govern its contracted decontamination work in the SDA are provided in the Common Specifications[28].

The decontamination technologies that have been institutionalized in the Decontamination Guidelines and the Common Specifications are primarily surface removal or washing and wiping methods. That is, few of the advanced decontamination technologies have been recommended by the MOE for full-scale use. This may be due to low performance in several areas, such as decontamination effectiveness, reliability, and waste volume; however, a cost-benefit evaluation process considering these areas and the waste management costs through final disposal does not appear to have been used in decision-making.

A report on decontamination effectiveness of the various decontamination methods that have been used was prepared by the MOE in January 2013[29]. The methods evaluated in that report are the typical methods cited in the

Decontamination Guidelines. Additional controls for quantifying decontamination effectiveness to reduce the observed variability are discussed below.

Each of the surfaces that will be encountered has a feasible decontamination technology/method that can be applied to it with the notable exception of forests. The vast expanse of forested regions makes it essentially impossible from a cost perspective to remove surface contamination and manage the waste for a method in which all the surface materials are removed.

In addition, the ESF team found potential general improvements that could be realized in efficiencies and cost- effectiveness in decontamination as recommended below. Additional methods used in the U.S. including the DOE complex may also provide ideas for improvements in efficiency and cost-effectiveness. Relevant documentation provided by the ESF team on this topic in contained in Appendix 3.

Decontamination Effectiveness

The decontamination effectiveness of a decontamination method applied to a surface is characterized in terms of the reduction of the radiation field above the surface. A compilation of decontamination methods for a variety of surface types, and the decontamination effectiveness is provided by the MOE[29].

The decontamination effectiveness, or reduction in contamination, in percent, is defined as[29]:

[1 - (*Count Rate After Decontamination*)/
(*Count Rate Before Decontamination*)] x 100

Both count rate meters (gas chamber design) and scintillation detectors (e.g. NaI and CsI) have been used, collimated and uncollimated, at surface contact, and at 1 cm, 50 cm, and 1 meter distances from the surface to measure the radiation fields from contaminated surfaces. However, no protocol has been formally established and is consistently used in establishing decontamination effectiveness.

The count rate measurements, as reported in reference 19, in most cases included the contribution from the background radiation. That is, shielded detectors were mostly not used in reporting the results and the radiation count rate was confounded by the background counts.

In addition, some of the data plots were confounded with more than one type of material of construction (e.g. roof of cement or metal) of the host material being decontaminated. Furthermore, the details of the application of the decontamination method and its control (e.g. sweep rate of the high-pressure water jet; the specific wipe media used to perform the wiping) were not well-specified.

In regard to the material of construction, other characteristics of the initial surface or host materials were not described, and there may be important features affecting the ability to decontaminate the material. For example, on a roof material, surface cracking may occlude water from reaching the cesium, and further, there may be clay in the cracks that chemically-bind cesium. Surface cracking of a roof tile, and clay debris in the tile cracks, may therefore make decontamination difficult. These distinguishing features should be noted to reduce the uncertainty in evaluation of decontamination effectiveness for a decontamination method applied to a surface.

Figure 3-1 below, reproduced from reference 22 shows an example of the decontamination effectiveness results for an asphalt road surface that had a range of initial contamination levels. It is readily apparent that there is a high variability in the decontamination effectiveness.

Influence of Remote Contamination on Air Dose Rate above a Surface

To demonstrate the need to use a collimated, shielded detection system to measure the radiation from over a surface of interest to evaluate decontamination effectiveness, the following two hypothetical cases are considered. Both of these cases show the potential impact to a dose rate measurement using an uncollimated, unshielded radiation meter that is made above a surface being decontaminated and adjacent to these "interfering" surfaces.

The first case is the effective dose rate contribution from a source near the surface of interest for decontamination. The source is a finite surface (10 m x 10 m) "contaminated" with Cs-137 (Ba-137m) such that the effective dose rate from the 0.662 MeV gamma radiation is 16 μSv/hr on contact. This interfering surface is perpendicular to the surface of interest being decontaminated. This first case is used to estimate the influence of a nearby contaminated building when evaluating decontamination of the ground surface next to the building.

The second case considers the effective dose rate contribution by a contaminated forested region (1 km x 1 km) with trees 10 m tall adjacent to the surface of interest (e.g. road) for decontamination. The radiation profile or source strength was set to a Gaussian distribution that varies with elevation such that the maximum source concentration is at ground level and is 16 μSv/hr at the forest/road interface (Figure 3-2). Less dense woods receive a smaller FWHM corresponding to a higher concentration of contamination near the ground and brush level. The detector is an uncollimated point detector in each case.

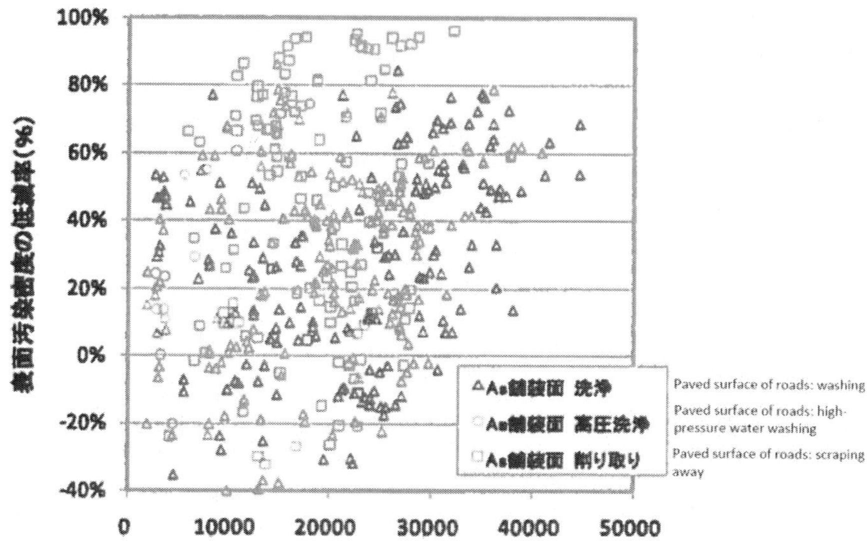

Figure 3-1. Decontamination effectiveness for contamination removal from asphalt road surfaces.

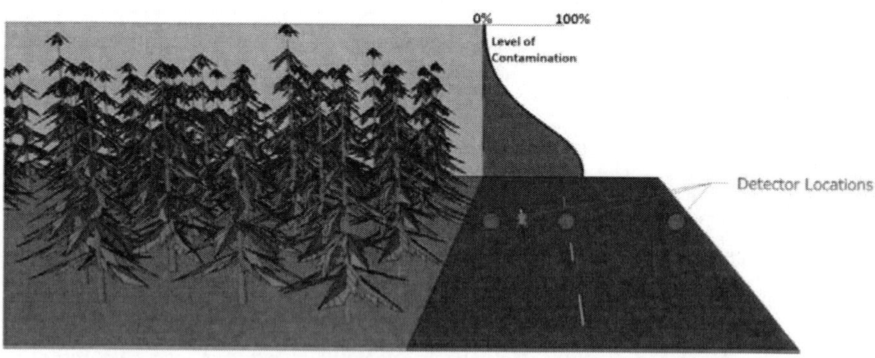

Figure 3-2. Assumed radiation profile from a forest region with a Gaussian distribution of contamination from the forest floor to the top of tall (10 meter) trees.

In the first case, if a dose rate measurement is being made above a road bed adjacent to a highly contaminated building 10 meters away, for example, Figure 3-3 shows that a significant dose rate from the building would interfere with the dose rate measurement being made on the road. At a distance of 50 meters, this interfering dose rate would be much reduced.

For a stretch of road, for example, next to a large heavily wooded forest (15% wood by volume), the results in Figure 3-4 show that distances up to 50 meters may be significant to the dose rate above the road if the forest is highly contaminated.

Furthermore, a contaminated patch of land with light forest (5% wood by volume) or with grassland (modeled as flat plane) can influence dose measurements being taken of a nearby decontaminated stretch of road. Figure 3-4 illustrates that for grassland foliage areas where cesium is concentrated near the surface with little wood mass to serve as shielding between the source and the uncollimated detector, the contribution from decontaminated regions can appreciably affect readings up to 100 meters from the road/grass interface. The effect is somewhat mitigated in heavily wooded forests (15% wood by volume) due to the more even distribution of cesium and the abundance of dense tree matter to serve as shielding.

These simple cases were meant to illustrate the point of potential interference in making dose rate measurements with an uncollimated detector to evaluate decontamination effectiveness.

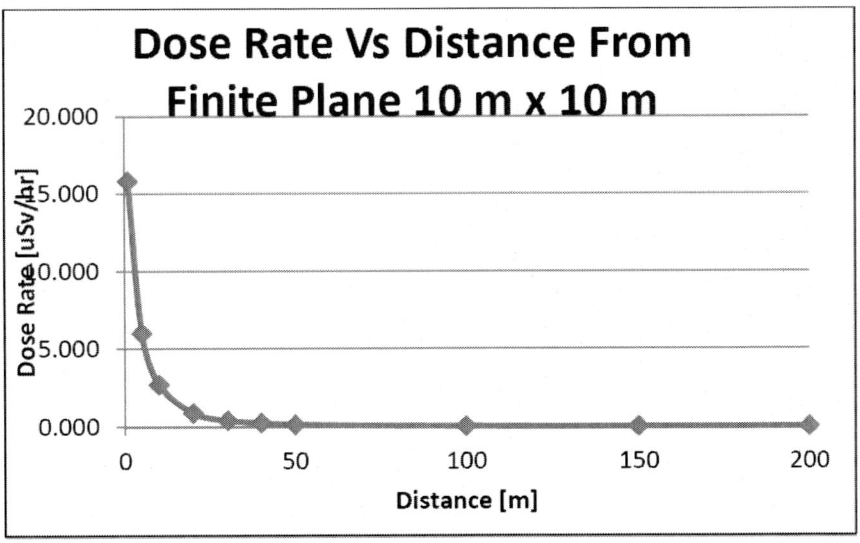

Figure 3-3. The effective dose rate as a function of distance from finite plane perpendicular to the surface of interest with radiation through an air medium [results using MCNP modeling by C. Verst, SRNL].

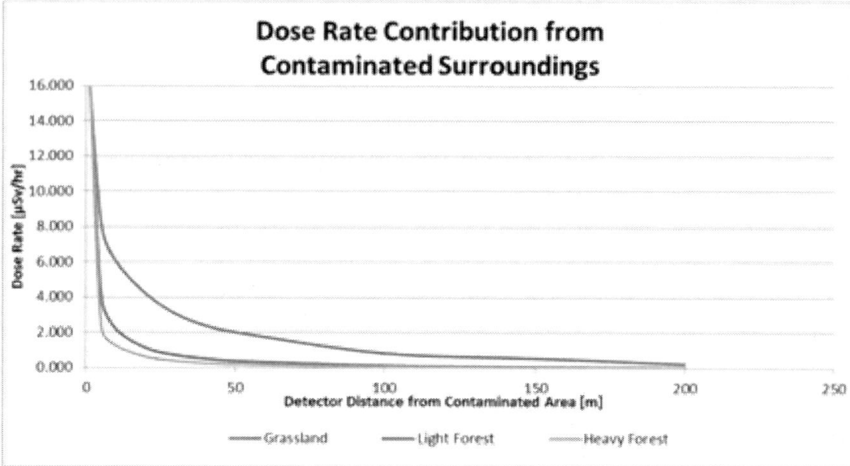

Figure 3-4. The effective dose rate as a function of distance from nearby foliage with radiation through an air medium [results using MCNP modeling by C. Verst, SRNL].

3.3. Recommendations and Supporting Actions

Recommendation 1

Develop and ensure application of a set of standard protocols for measuring the effectiveness of decontamination methods (before-after) for all applicable targets of decontamination (e.g., roads, soil, etc.)

The primary need for critical evaluation of the removal of the radiological contaminants from the infrastructure and land surfaces is a field dose rate measurement protocol before and after decontamination to establish decontamination effectiveness.

Decontamination effectiveness for removal of cesium from surfaces and the sub-surfaces needs explicit definition and controls in its measurement for two primary reasons:

1. Enable critical review from a controlled-method comparison of decontamination technologies/methods applied to a surface. A catalog of with a complete description of the surface; decontamination method; and decontamination effectiveness would enable well-founded down-selection of a decontamination technology and

application method for implementation for full-scale decontamination work

2. Provide a system for a field check of the decontamination work to ensure consistency and thoroughness in a decontamination project

Further recommendations for specification of the protocol are not provided here. Shielded, collimated detector configurations have been used for some of the work, and they provide a good starting point to adopt for the radiation measurement. A team of radiation measurement experts would be expected to develop the protocol. Also, it is recognized that several versions of the protocols, for soils, road surfaces, and grass fields, for example, may be needed.

It is highlighted that uncollimated radiation monitoring to establish the effective dose rate at a location is a separate and critical need for radiation protection purposes. That is, using an uncollimated detector to characterize the effective dose rate at a height of 1 meter and 50 cm above the ground is a good approach to provide information to evaluate the effective whole body dose rate to the human adult and child, respectively. This has been the standard approach to date to characterize the effective dose rate from areas in the contaminated regions at a point location as the primary information for consideration of annual dose estimation to the public. The characterization of the effective dose rate at a height of 1 meter before and after decontamination of a surface has also been used in the work to date to characterize the decontamination effectiveness. But, as shown in the example hypothetical cases above, the background dose rate from such measurements can interfere with the evaluation of decontamination effectiveness.

Supporting Actions

- Provide guidance for consistent and systematic application of the protocols to the decontamination efforts that are underway in: (a) local communities, (b) MOE-led decontamination activities in the evacuation zones, and (c) MOE-led efforts in high-dose areas.
- Provide contractual guidance to require implementation of these protocols and reporting of relevant data.
- Collect and organize decontamination performance data that are generated under the protocol in a manner that supports systematic analysis of the performance of the methods used for the full range of surfaces and land conditions.

Recommendation 2

Conduct a systematic analysis of the existing performance data to identify potential factors or practices that could improve effectiveness of future decontamination efforts and that identifies situations where specific practices are not likely to be effective.

This recommendation can be implemented immediately as part of the initial work to compile, evaluate, and prepare the information for the catalog described in recommendation 3.

Supporting Actions

- Identify problematic surfaces or material types that can identify priorities in recommendation 4.
- Provide technology performance data to support recommendation 3.

Recommendation 3

Develop and maintain a comprehensive catalog of decontamination technology performance (based on systematic methods for assessing effectiveness).

As described in the observations, methods to measure the radiation level before and after decontamination have varied, and a single field protocol has not been prepared or consistently implemented for the decontamination work to date.

There is a need to prepare a catalog with detailed, explicit descriptions of the host surfaces, decontamination method and application controls, and the field protocol to establish the dose rate before and after decontamination. Of special note is that soils and other surfaces may have a large depth profile of up to several centimeters from the top surface position. It would be useful to have this profile characterized as part of the information set.

There are 3 parts of applying a decontamination technology with an application method and evaluation of the decontamination effectiveness. Each of these parts is expected to be well-described to critically evaluate decontamination effectiveness of a decontamination method in terms of the reduction of dose rate above a surface:

1. Full description of the surface (e.g. to distinguish between different types of roofing materials and its physical condition);

2. Specification to control the decontamination method (e.g. water spray pressure, the distance of nozzle from the surface, rate of travel of the spray head across the surface); and
3. Field protocol (standard) for measurement of the dose rate from the surface (no background) before and after decontamination.

The information in the catalog would be compiled with a radiation dose rate collected with the field protocol and with the surfaces and decontamination methods well-described.

Supporting Actions

- Include key performance characteristics related to implementing each method including effectiveness, cost, materials, waste generation, etc.

Recommendation 4

Enhance existing processes for facilitating the development and maturing advanced decontamination technologies.

There have been several special campaigns to investigate advanced decontamination technologies. Some of the technologies have been adopted and are part of the Decontamination Guidelines[27]. However, some of the technologies have not been developed further. The adaptation of a new technology and/or method for application is driven by the potential savings in cost and effectiveness of decontamination or both. The process to mature the technology and adopt the technology should implement the recommendations 1 and 3 to control the process, and should look at the full remediation system, including such considerations as cost savings and risk avoidance in handling, transportation, storage, and ultimate disposal of waste that would result with a reduction of waste volume from the advanced technology, for example.

Supporting Actions

- Develop a government-provided test bed for advanced decontamination technology for the more intractable decontamination challenges such as forest lands and agricultural lands. The test-bed provides a readily available area with representative contamination conditions that can be available to companies, universities, laboratories, or others offering innovative or experimental methods.
- Include contract incentives for soliciting advanced decontamination technologies.

- Use the to-be established field decontamination effectiveness protocol in recommendation 1 to evaluate new technologies.

4.0. Assessment of Remediation System Program Elements: Waste Management System

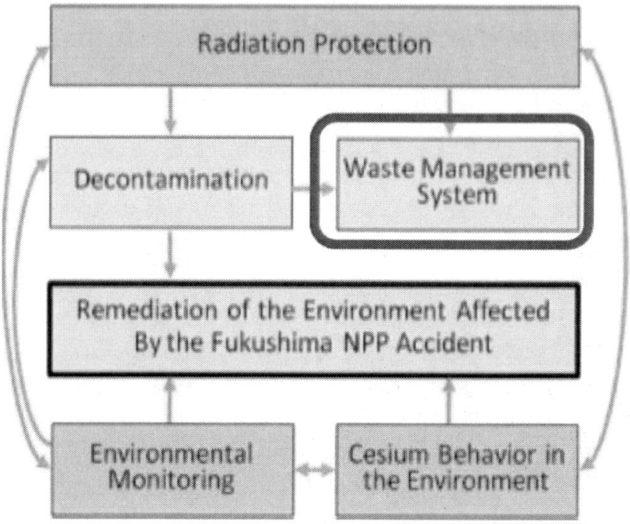

Waste management is a major challenge for the overall remediation of the area affected by the Fukushima NPP accident. Two years after the accident a very large portion[30] of the waste generated by decontamination efforts still sits at the widely distributed locations where it was generated. There is broadly varied progress in implementing "temporary storage facilities" (TSFs) in the communities undergoing decontamination efforts. There is an urgent need to put in place all of the elements of an integrated waste management system to safely gather and manage the large volumes of waste that are being generated. Elements of an integrated waste management system include:

- transportation
- treatment
- storage
- disposal.

4.1. Current Situation

Overview of Waste Management Policies

Part 3 of the Act on Special Measures (2011), "Basic matters concerning the disposal of waste contaminated with radioactive materials discharged by the accident," defines the national policy on the management of "Specified Waste" – materials that were contaminated by radioactive materials discharged by the accident[31]. Part 5 of the Act on Special Measures (2011), "Basic matters concerning the collection, transfer, storage and disposal of the removed soil," defines the national policy for managing the waste generated by decontamination activities, particularly soil.[32] Figure 4-1 provides an overview of the managing both specified waste and decontamination waste. This chart is specific to Fukushima Prefecture[33]. Because of the limited time available for ESF review, the emphasis for the review and recommendations is on the systems and processes for handling soil and waste generated as a result of decontamination work in Fukushima Prefecture.

Waste Storage

There are three types of storage areas planned for decontamination waste in Fukushima Prefecture: on-site storage, temporary storage facilities, and interim storage facilities.

"On-site storage"– This term refers to storage of decontamination waste at its point of generation, including residential, commercial, agricultural, and other public facilities. This waste is stored at the point of generation pending availability of a temporary storage facility. On-site storage is the responsibility of each municipality in the ICSA. Within the SDA, on-site storage is associated with various demonstration projects that are underway by MOE. Full-scale decontamination efforts cannot begin in the SDA until Temporary Storage Facilities are available.

- Guidelines for safe storage practices are provided in "Decontamination Guidelines, Part 4: Guidelines Pertaining to the Storage of Removed Soil".[34] Various methods for containing decontamination waste are described with the intent to ensure that collected materials are contained from the elements, prevented from release, stored in a manner to minimize external dose, and segregated from different types of waste (e.g., soil, burnable debris, etc.). The ESFs frequently observed storage method was the use of flexible container bags (see Figure 4-2) that prevent the waste from becoming airborne or dispersing and that are strong enough to be lifted with failing.

- In Fukushima City, the ESFs witnessed decontamination work directed by the city government in a residential area. Because Fukushima city does not have sufficient TSF capacity, residential decontamination waste must remain at the residence and may either be stored above ground on the property pending availability of the municipal TSF or buried on the property (with markings). City officials indicated that 80% of residents choose burial on their property. The left three pictures in Figure 4-3 show three areas with on-site storage that were observed: Naraha Town agricultural fields, Tomioka Town Sports Complex, and Fukushima City private residence.

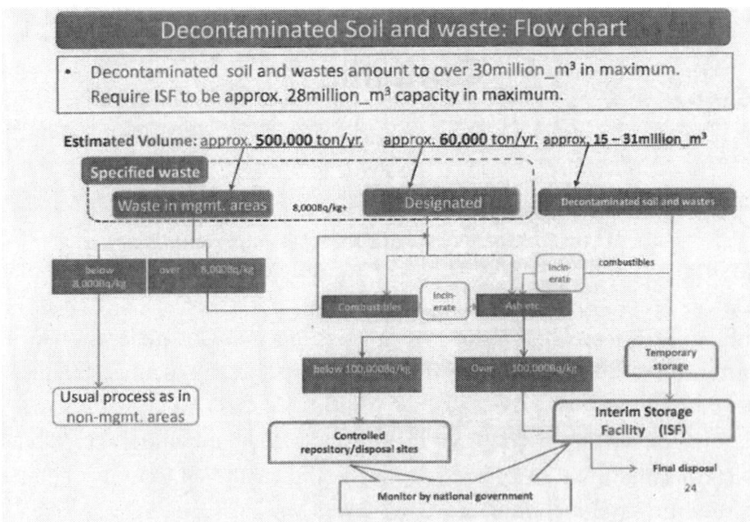

Figure 4-1. Logic for managing specified waste and decontamination waste for Fukushima Prefecture.

- Fukushima Prefecture provided monthly summaries of on-site and TSF storage facilities (number of locations but not quantity of stored waste). For the communities outside of the Special Decontamination Area, as of December 2012, there were nearly 5,000 on-site storage locations in 52 communities. This includes nearly 3,000 private residences and commercial businesses, more than 1000 schools, kindergartens, and childcare facilities, and nearly 900 parks.[35].

CESIUM CONCENTRATION IN DECONTAMINATION WASTE

The concentration of cesium in decontamination waste will vary depending on the decontamination methods applied and the surface or medium being cleaned. As a point of reference, however, MOE has provided the following information in the Decontamination Guidelines:

“The radioactivity concentration of soil at an air dose rate of 3.84 μSv/hour equal to 20 mSv/year is about 39,000 Bq/kg, which is obtained by converting soil monitoring data given by the Ministry of Education, Culture, Sports, Science and Technology as well as soil monitoring results including air dose rates obtained at elementary schools in Fukushima Prefecture to the value on June 1, 2011, and using the resulting regression formula “Log (Air dose rate) = 0.815 × Log (Cs concentration)   3.16” (Japan Nuclear Energy Safety Organization).”

Japan’s radioactive waste disposal requirements place an upper concentration limit of 100,000 Bq/kg for shallow trench disposal. Greater than 100,000 Bq/kg requires a more rigorous disposal using a shallow pit method.

In its present form, i.e., without concentration, most of the decontamination waste would meet current requirements for the least restrictive form of low-level waste disposal. However, these regulations were not developed in anticipation of this extremely high volume waste type.

Temporary storage – This term refers to new storage facilities that are prepared in communities undergoing full- scale decontamination efforts. The intent is to move waste from these facilities to longer-term interim storage facilities within 3 years of startup of temporary storage. TSFs are the responsibility of the municipalities in the ICSA and the responsibility of the national government (MOE) in the SDA. TSF requirements and conceptual designs are provided in “Decontamination Guidelines, Part 4: Guidelines Pertaining to the Storage of Removed Soil.”

- Outside of the Special Decontamination Areas, local communities are to secure the locations for their own facilities. Inside the SDA, MOE will secure the locations with agreement of the community government and citizens.
- According to "Status of Decontamination Storage for Fukushima Prefecture," 475 TSFs were identified in the 52 communities outside of the SDA. These vary in size and configuration and some municipalities still do not have an existing TSF and others do not have sufficient capacity to yet store newly generated decontamination waste. Consequently, this lack of capacity leads to on-site storage.
- There is a fundamental policy difference between areas outside the SDA and inside the SDA regarding necessity for having temporary storage. Inside the SDA, MOE cannot begin full-scale decontamination work until TSF capacity is secured. Outside of the SDA, the communities were able to initiate decontamination work due to the urgency of cleanup of the environment to protect residents. But after two years, many communities still lack the necessary TSF capacity and must rely on "on-site" storage.

Figure 4-2. Flexible Container Bag for Decontamination Waste.

- Inside the SDA, as of December 2012, MOE has worked with four cities to secure TSFs to enable full scale decontamination work to begin: Tamura city, Naraha town, Kawauchi village, and Iitate village. TSFs are partly secured for Katsurao village and Kawamata town and the process to secure facilities and begin full- scale decontamination is underway in the remaining five towns.

Interim storage – This refers to the large storage facilities for decontamination waste and soil and for designated waste that may exceed 100,000 Bq/Kg. These facilities are planned only for Fukushima prefecture. The goal is to begin operation of one portion of ISF capacity by early 2015.

- Interim storage facilities (ISFs) are to be built by the Ministry of the Environment. At present, ISFs are only planned for Fukushima Prefecture as the expected volumes of decontamination waste and soil are much smaller in other prefectures.
- The current plan is to build ISFs in close proximity to the Fukushima Daiichi and Daini NPPs. Facilities would be located in Naraha town, Okuma town, and Futaba town.
- The estimated volume of waste to be handled by the ISFs is between 15 million m^3 to 28 million m^3 but this estimate has not been updated since it was first produced for the October 2011 “Basic Policy on Interim Storage.” The “Basic Policy on Interim Storage” estimates the total area required to be approximately 3 km^2 to 5 km^2.[36]
- Figure 4-4 shows the schedule and roadmap for deploying ISFs. The selection of sites still has not occurred so the actual status of that step is roughly a year behind this original plan. MOE staffs are currently working steps 1 through 4 in parallel in an attempt to meet the goal of initial operation in early 2015.
- A significant challenge for site selection for the candidate locations in these three towns is the need to purchase land from many private land owners. Because these are evacuated areas, it is difficult to find the land owners and commence negotiations. Moreover, there are differing views among the land owners about whether they want to return to their land or not, and land purchase decisions are linked in the land owners’ minds with compensation due to the accident.

Figure 4-3. Photographs of Various Decontamination Waste Storage Sites in Fukushima Prefecture.

Waste Treatment

Policy statements related to waste treatment for decontamination waste and soil:

- From the Act on Special Measures, Part 5: "Meanwhile, from perspectives to secure capacity of the temporary storage facilities, etc., an attempt shall be made during the storage or the disposal process ***to minimize volume of the removed soil capitalizing on technological developments***. At the same time, while ensuring safety and security, recycling options for the removed soil whose levels of contamination is low, such as that separated as a result of volume reductions, needs to be considered." (Emphasis added.)
- From the Basic Policy on Interim Storage, Part 3: "An intermediate treatment system will be developed on the premises of a facility where this is technically feasible and will introduce effective treatment methods to help with safety management and volume reductions."
- From the Basic Policy on Interim Storage, Part 4: "Since the final disposal methods are largely dependent on progress in the development of technologies, including the effective separation and concentration of radioactive materials, the national government shall endeavor to conduct relevant technological research and development and assessment."

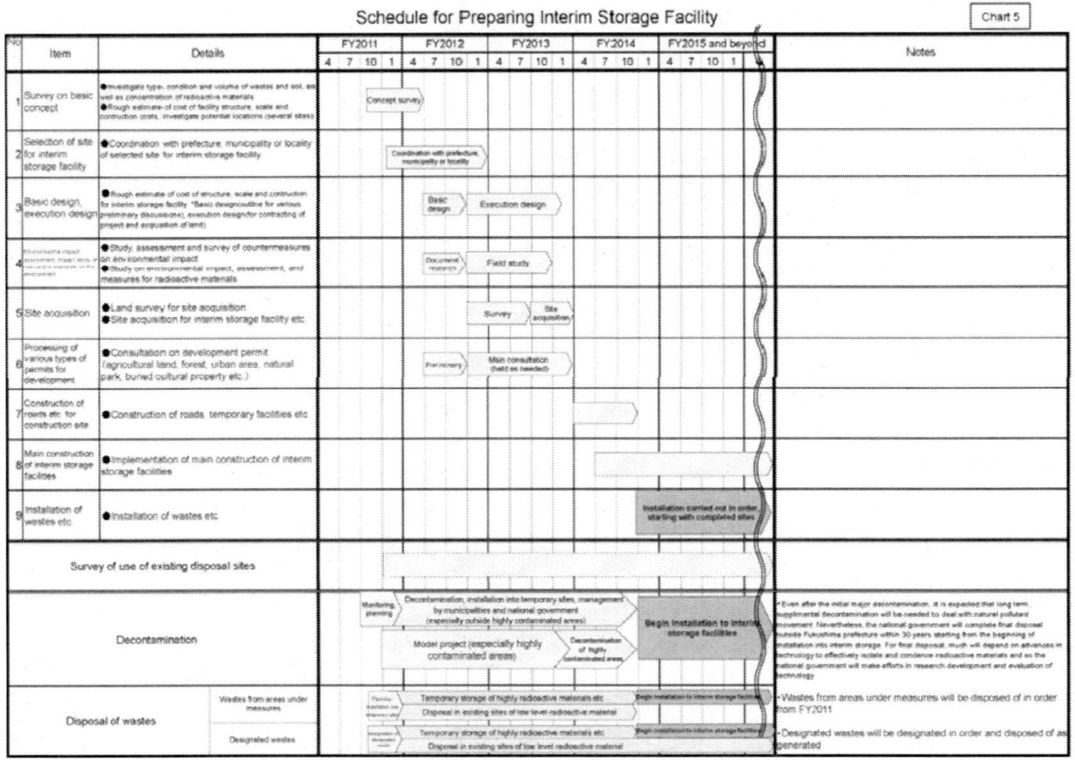

* For decontamination, installation into interim storage facilities will start after 3 years from beginning of the installation into temporary facilities.

Figure 4-4. Schedule for Deployment of Interim Storage Facilities in Fukushima Prefecture.

Waste Disposal

The policy on waste disposal is that "the national government shall complete the final disposal outside Fukushima prefecture within 30 years from the start of the interim storage" (From the Basic Policy on Interim Storage, Part 4).

Waste Transportation

Very little information on transportation was reviewed during the ESF assignment. The primary focus for transportation appears to be related to transferring decontamination waste from TSFs to the planned Interim Storage Facilities. As shown in Figure 4-5, consideration of transportation routes within Fukushima Prefecture led to the determination that three separate ISFs should be developed to serve different portions of Fukushima Prefecture as shown on the map.

If there is 28 million m^3 of decontamination waste that needs to be moved and each truck is assumed to hold 10 m^3, then 2.8 million truck shipments could be required[37].

The national government has prepared reasonable guidance for temporary storage facilities and the designs that were seen seem quite robust for the planned mission. But, MOE staff indicated that there is significant public opposition to site selection of the needed TSFs in both the ICSA and in the SDA.

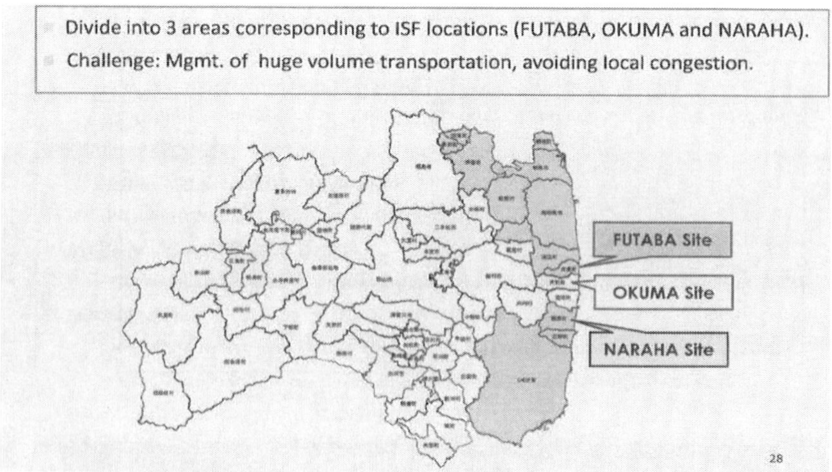

Figure 4-5. Transportation consideration led to plan to develop three Interim Storage Facility locations serving different portions of Fukushima Prefecture.

4.2. Summary of Observations

Figure 4-6 illustrates the required elements of an integrated waste management system that would handle the large volumes of decontamination waste that are being generated. Japan currently lacks many of these elements. Key observations include:

- Significant delays are being experienced in the development of temporary and interim storage facilities needed to contain the decontamination waste that is being generated. These delays lead to the proliferation of waste storage at decontamination waste site. This prevents visible progress on overall remediation from being made.
- Effective volume reduction methods for some categories of waste are available (e.g., incineration) but there is substantial public opposition. For example, there is a perception that incinerators will release cesium from the off gas system. Consequently, the needed capacity is not available leading to generation and storage of unnecessary volumes of remediation waste.
- There is no explicit decision process to evaluate "system wide" treatment methods (e.g., soil washing, incineration, volume reduction, segregation, recycling/reuse) for decontamination waste. National policy statements clearly recognize the importance of volume reduction and treatment of decontamination waste, but ESFs did not find evidence of any systematic evaluation of treatment options.
- The estimated volume of decontamination waste to be generated in Fukushima Prefecture has not been updated since 2011 and may not accurately reflect the results of completed and ongoing decontamination work.
- There is a lack of an overall inventory or rollup of decontamination waste generation and storage. Some of the information on storage locations in municipalities is not publicized to prevent illegal dumping.
- The transportation of the huge volume of decontamination waste from thousands of locations to ISFs is likely to be difficult challenge that will need to be faced even before the ISFs are ready to begin receiving waste.

4.3. Recommendations and Supporting Actions

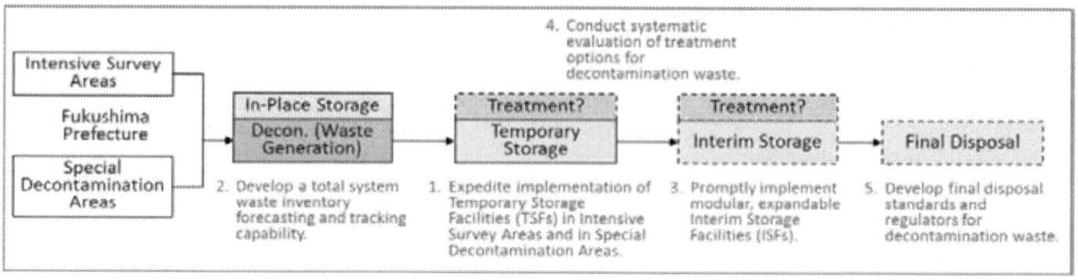

Recommendation 1

Expedite implementation of Temporary Storage Facilities (TSFs) in Intensive Contamination Survey Area and in Special Decontamination Area.

Perhaps the most striking reminder of the widespread decontamination efforts that are underway are the ubiquitous bags of decontamination waste that remain at the sites where the waste was generated. The lack of temporary storage facilities in many communities means that decontamination waste must be stored on the site where it was generated. The lack of temporary storage facilities has also prevented the startup of full-scale decontamination work in more than half of the communities within the Special Decontamination Area. Within the ICSA it should not be acceptable to continue to generating decontamination waste without available temporary storage. Temporary storage areas are necessary to alleviate continued accumulation of decontamination waste at generation sites. TSFs are needed to remove and consolidate waste from widespread on-site storage locations.

Supporting Actions

- Encourage selection of TSFs prior to large scale decontamination work; one possible pre-condition for proceeding with decontamination work in any given community can be that community's selection of a fully compliant TSF to receive the material.
- Increase communication with local governments and showcase successful TSFs. Consider empowering community/stakeholder working groups to provide advice on the holistic remediation AND waste management approaches for each community.
- Consider use of nationally-owned or publicly-owned land (e.g., contaminated forest land). Some communities have been far more successful than others in gaining public approval for TSFs. Those

success stories need to be highlighted and used as examples for other communities to follow. For some communities without established TSFs, options for using publicly-owned land should be considered. One option would be to set aside forest land, for which decontamination is not planned, for use as temporary storage.

Recommendation 2

Develop a waste inventory forecasting and tracking capability that incorporates a systems approach.

The key to systematic planning and design of a waste management system is an accurate base of information that describes the quantities and characteristics of the waste that has been generated and that will be generated. There are many elements of this important data set that already exist, but it is important to maintain a disciplined process for collecting the relevant data. Typically, waste forecasting tools are built around a set of assumptions. These assumptions would be replaced over time by actual observed waste generation rates and waste characteristics. This information is essential to planning and optimizing the treatment, storage, and disposal system components.

Supporting Actions

- Provide a template for collecting a comprehensive data set from each municipality: volume by type of waste (soil, combustible, non-combustible, etc.); location or facility type; surface dose rates for bags or containers; and activity concentrations (Bq/kg).
- Use the forecasting tool to conduct options analyses to support analysis of high-level remediation strategies and priorities, including assessment of treatment, storage and transportation options and impacts.

Waste forecasting tools are useful for assessing the impacts of changes in assumptions or strategies. For example, changes in preferred soil removal depth can be evaluated in terms of the impacts on the total waste to be generated. One possible issue that could be evaluated would be the impact on total waste volume if the bags of clean soil used for shielding on the perimeter of storage areas would need to be sent to Interim Storage Facilities. How much additional storage space would be required? Also, alternative waste accumulation and transportation strategies can be evaluated.

Recommendation 3

Promptly implement modular, expandable Interim Storage Facilities (ISFs).

Progress toward establishing ISFs could alleviate one obstacle to selecting TSF locations in some communities. While the GOJ's plan is to implement three interim storage facilities with sufficient capacity to accept the total expected volume of decontamination waste by early 2015. There are numerous obstacles to achieving this goal. But even if the ISFs are constructed as planned, the logistics of moving 30 million m^3 of bags of decontamination waste will take a very long time. Consequently, a modular/expandable design would preserve options to optimize volume reduction or other treatment processes.

Supporting Actions

- Complete surveys/investigations of candidate sites in Futaba, Okuma and Naraha towns.
- Develop design concepts that balance the need to start receiving waste as soon as possible (e.g., by the target date of January 2015) but also allow evaluation of treatment methods (e.g., volume reduction and/or stabilization) that could reduce total storage requirements and produce more robust waste forms for final future disposal.
- Design ISF to have expandable modules compatible with the land topography of the selected locations.
- Design ISF to have areas for testing, demonstration and implementation of treatment methods (e.g., volume reduction).
- Accommodate treatment decision from Recommendation #4.
- Complete the facility design and prepare an evaluation of ISF options to allow for public review and comment on the ISF site selection and technical approach.
- Update the ISF implementation schedule based on an assessment of the current status and remaining actions for deployment.

Recommendation 4

Conduct systematic evaluation of treatment options for stabilization and/or volume reduction of decontamination waste.

There is extensive pressure to obtain and start using an ISF as soon as possible to alleviate the accumulation of decontamination waste in widespread communities. This pressure has been cited for the need to defer evaluation of eventual volume reduction methods for various types of decontamination waste. But, the evaluation of treatment options needs to also consider the potential for solutions that could be implemented on a distributed basis at temporary or on-site locations. Moreover, evaluation and selection of treatment options could conceivably reduce the total storage volume required for the ISFs. An explicit and visible decision process could be used to evaluate a wide range of options including no treatment, soil washing, incineration, separations methods, thermal treatment methods and other methods that technology providers could offer. In addition, the viability and requirements associated with reuse of "cleaned" soil should be part of this evaluation. Some methods such as incineration may be less acceptable to the public because of a perceived potential to release cesium from offgas systems. A full range of evaluation criteria should be considered including cost, technical performance, volume, reduction, maturity, final waste form, and public acceptability.

Supporting Actions

- Identify candidate concepts for testing and implementing large scale treatment systems (e.g., volume reduction) for decontamination waste.
- Provide a systematic technical, cost, and benefit evaluation of candidate treatment options that reflect a full range of potential options (no action, incineration, soil washing with reuse, thermal methods and distributed versus centralized methods).
- Publish results with preferred alternative for public review and comment.

- Place contracts to design, build and operate and consider subsequent step to evaluate treatment options.

Recommendation 5

Develop final disposal standards and regulations for decontamination waste.

Final disposal of decontamination waste is not a current topic of discussion. This is typically viewed as something that will be dealt with much later. Current plans call for storage of decontamination waste for up to 30 years in the ISFs followed by final disposal somewhere outside of Fukushima prefecture. But, definition of final disposal standards and assessment of potential methods and locations is an important factor in optimizing the waste treatment approach. Consequently, it is important to begin defining the final disposal or end-state requirements for decontamination waste to support design of a rational storage and treatment system.

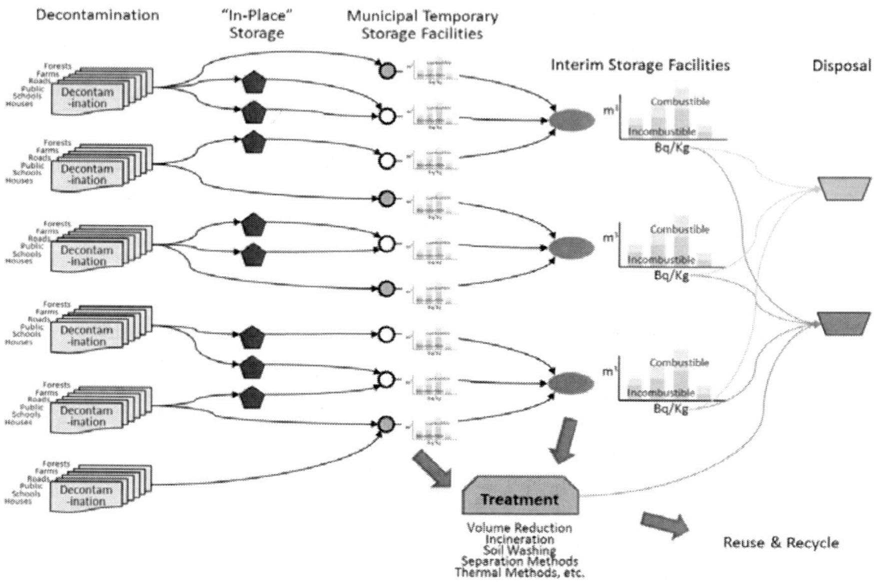

Figure 4-6. Conceptual Diagram of an Integrated Waste Management System.

Supporting Actions

- Evaluate estimated characteristics of decontamination waste categories relative to existing Japanese and international radioactive waste regulations or guidelines.
- If necessary, develop disposal system and waste form requirements based on a performance assessment using acceptable international standards and methods.
- For selected waste categories (e.g., contaminated soil, trees, etc.) evaluate the potential costs and benefits of coordinating some elements of off-site disposal with Daiichi decommissioning and remediation waste materials generated on site.

5.0. Assessment of Remediation System Program Elements: Environmental Monitoring

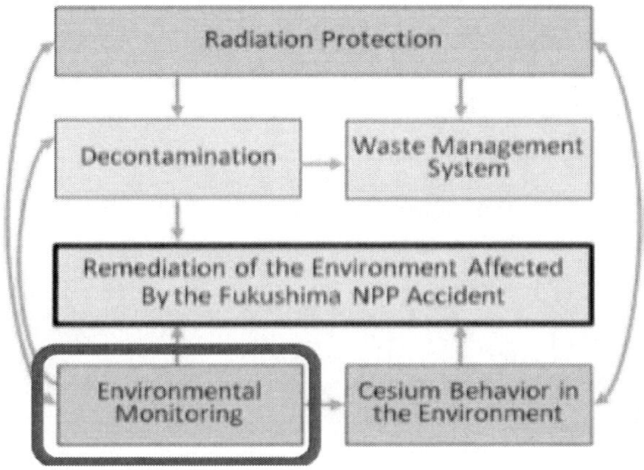

Environmental monitoring supports all other elements of the remediation system. Environmental monitoring is essential for providing feedback on the distribution of contaminants resulting from the accident and on the progress of remediation efforts. Monitoring data is essential for making contaminant transport models more representative of actual behavior. Finally, environmental monitoring should support development of radiation protection strategies for people living in areas with long-term contamination from the accident.

5.1. Current Situation

There are multiple purposes for environmental monitoring related to the Fukushima accident. During the course of the accident, monitoring was focused on understanding the spread of the plume from the accident to support population evacuation and protection decisions. Once the threat of additional releases ended, monitoring focused on assessment of the extent of contamination, identification of pathways for human exposure and using that information to ensure that appropriate measures are taken to protect the public. This section describes the current efforts that are underway by multiple agencies to support comprehensive radiation monitoring related to the Fukushima NPP accident.

Reform of Japan's Nuclear Regulation Organizations and Systems[38]

The Japanese government passed an act to reform nuclear regulation responsibilities on June 27, 2012. The Nuclear Regulation Authority (NRA) was formed in September 2012. The reform efforts separate the nuclear regulation function from the nuclear promotion function. NRA was formed as an independent organization but is affiliated with the Ministry of the Environment. NRA was formed to integrate nuclear regulation functions including nuclear safety, security, safeguards, radiation monitoring, and regulation of radioisotopes.

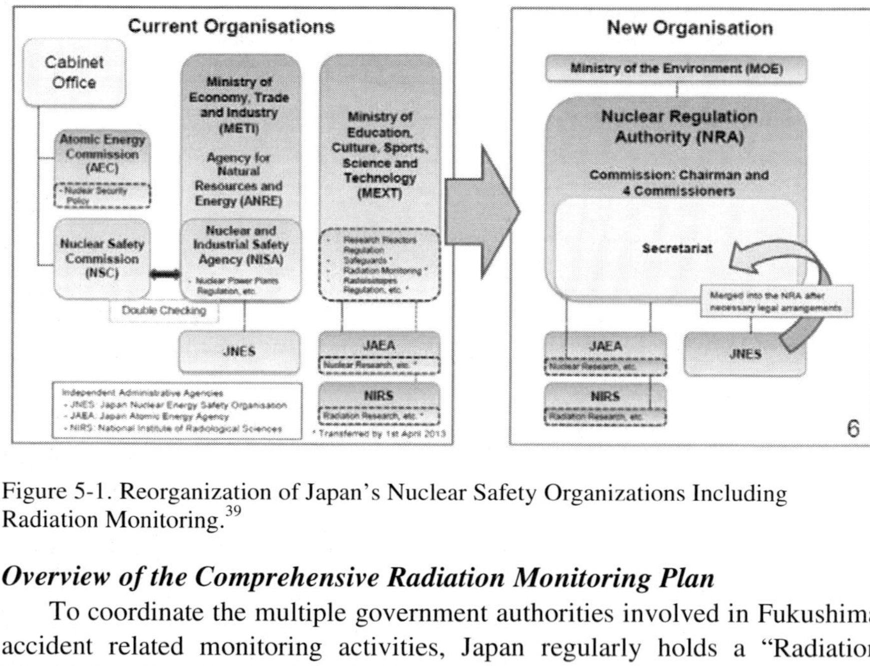

Figure 5-1. Reorganization of Japan's Nuclear Safety Organizations Including Radiation Monitoring.[39]

Overview of the Comprehensive Radiation Monitoring Plan

To coordinate the multiple government authorities involved in Fukushima accident related monitoring activities, Japan regularly holds a "Radiation Monitoring Coordination Meeting" that produces a *Comprehensive Radiation Monitoring Plan*[40]. Japan's Ministry of Education, Culture, Sports, Science and Technology (MEXT) served as the "control tower for total coordination and information aggregation." This function, however, was transferred to the Nuclear Regulation Authority (NRA) through its implementing legislation and this transfer of roles took effect at the start of FY 2013 (April 1, 2013).[41] Many of the graphics and materials provided in this section do not reflect this change in responsibilities.

The *Comprehensive Radiation Monitoring Plan* lists the following major objectives:

(i) Understanding of the distribution of radiation doses and radioactive substances mainly in areas and places where people reside on a mid- and long-term basis
(ii) Estimation of current exposure (external and internal exposure) doses of people living in the affected regions and their potential exposure doses in the future
(iii) Consideration and planning of measures for reducing exposure doses, such as decontamination, in accordance with various circumstances.[42]
(iv) Consideration and judgment for changing or reviewing the designation of areas under evacuation order through estimating future exposure as realistically as possible
(v) Acquiring of basic data for managing residents' health and assessing effects on their health
(vi) Understanding of the dispersion, deposition, and migrations of radioactive substances released in the environment.

The *Comprehensive Radiation Monitoring Plan* also defines the roles for conducting monitoring and for ensuring close collaboration among related ministries and agencies, local governments, and the nuclear operator and related company. This allocation of roles as of March 1, 2012 was the following:

- MEXT: Serving as the control tower for total coordination and information aggregation; Carrying out environmental radiation monitoring
- Nuclear Safety Commission of Japan: Giving advice to related ministries and agencies; Comprehensively assessing the measurements and the analysis of measurement results carried out in monitoring conducted by related ministries and agencies
- Nuclear Emergency Response Headquarters (Local Nuclear Emergency Response Headquarters and Team in Charge of Assisting the Lives of Disaster Victims): Carrying out and coordinating radiation monitoring around TEPCO's Fukushima NPPs in cooperation with related ministries and agencies; Offering assistance to monitoring conducted by Fukushima prefecture
- Related ministries and agencies: Aggregating information on monitoring, offering assistance, and conducting analyses in line with administrative objectives

- Local governments: Carrying out community-based monitoring and transmitting information integrally, in collaboration with the government and the nuclear operator and related company
- Nuclear operator and related company: Under the initiative of the government, carrying out monitoring together with local governments and transmitting information integrally with the government Figure 5-2 provides a synopsis of the monitoring activities carried out in support of the comprehensive plan. At least nine national ministries and agencies are identified along with prefectural and local governments.[43]

Additional monitoring plans have been developed[44]

Monitoring Information Availability

The recently developed NRA monitoring website[45] includes a broad range of monitoring resources organized into the following areas:

- Monitoring of the environment in general
  - National monitoring
  - Terrestrial monitoring around Fukushima nuclear power plant
  - Throughout Fukushima Prefecture
  - Monitoring details for evacuation areas
- Marine monitoring
- Monitoring information for schools, etc.
- Monitoring ports, airports, parts, and sewer
- Environmental monitoring of water, natural parks, and waste
- Soil monitoring farmland, forest, and pasture
- Monitoring food and water
- Radiation dose distribution maps, etc.

One additional site with map based access to automated air dose rate monitoring information is available for Fukushima prefecture[46]

The recent creation of the Nuclear Regulation Authority not only reforms nuclear regulation in Japan but also assigns the overall coordination responsibility for Fukushima-related environmental monitoring to the NRA. This is a very positive step that can improve the accessibility of monitoring data and strengthen the integrated analysis of multiple data sets.

5.2. Summary of Observations

Japan has developed many elements of an effective long-term environmental monitoring program. Key observations include:

- Multiple monitoring data sets are being collected by national, prefectural and municipal government entities. More than 10 national agencies and many separate prefecture and municipal agencies are involved in collecting, analyzing and reporting monitoring data.
- There is a multi-agency "Radiation Monitoring Coordination" function that meets periodically to review and update the Comprehensive Radiation Monitoring Plan. This group provides an essential coordination function for the diverse group of agencies that have a role in collecting information relevant to the Fukushima accident impacts.
- While past monitoring data are available, much of the data are in the form of separate pdf files each containing a single image. Real-time air gamma dose monitoring station information is available through a map-based (GIS) viewer, although multiple levels of query information must be entered first to access any given map for a specific area.
- The monitoring plans made available to the ESFs describe the medium or target to be monitored, the methods for data collection, and the frequency. But, these plans do not describe how the data will be used for decision making purposes.

  This is a critical gap in building an effective long-term monitoring program. Specifying the decisions that need to be made allows one to also specify the desired quality and resolution of the data to be collected. By understanding beforehand the purpose for the monitoring data, it becomes much more likely that an effective long-term monitoring program will be built and maintained.

	Object to be Measured	Measuring Method	Location	Frequency	Competent Authority
Dose	Ambient Rate	Using Monitoring Post	675 points in and around Fukushima Prefecture 47 points at Prefectural Capital + 250 points in all over Japan 2700 Points at schools	Continuous	MEXT
		Using Survey Meter	55 Points within a 20km radius from 1F Site where disaster prevention countermeasures were provided, or in "Areas where Evacuation Orders are Ready to be Lifted", if necessary	Once a week	TEPCO
			35 points in Fukushima Prefecture for emergency monitoring	Twice a week	Fukushima Pref.
			3758 public facilities in Fukushima Prefecture	~100 facilities in every month	MEXT
			4 points in Tokyo Bay	Every day	MLIT
			2 points of airport(Narita, Haneda)	Every day	MLIT
			391 points of forest in Fukushima Prefecture	As needed	MAFF(Forestry Agency)
		Using Monitoring Car	143 fixed-points in Fukushima Prefecture	Once a three weeks	MEXT
		Airborne Monitoring	Area within an 80km radius from 1F Site	Once or twice a year	MEXT
	Accumulated Dose	Using Electronic Dosimeter	9 points out of a 20km radius from 1F Site 6 points within a 20km radius from 1F Site	Continuous (From end of the March, 2011)	MEXT FEPC
Radioactivity	Particles in an Atmosphere	Sample Analysis	14 points in Fukushima Prefecture	Continuous (1 point) Once a month(Others)	MEXT
	Settled Particles	Sample Analysis	Prefectural Capital , 1 point was added in Fukushima Prefecture after the 1F accident	Once a day(Added point) Once a month (Others)	MEXT
	Soil	Sample Analysis	4 points in Fukushima Prefecture	Once a three months	MEXT
			3423 points of farmland, field, orchard, grassland or pasture in Fukushima Prefecture	Once a year	MAFF
			391 points of forest in Fukushima Prefecture	As needed	MAFF(Forestry Agency)
	Sea Water	Sample Analysis	~60 points in coastal sea area (within 30km from coast)	Once a month	MEXT, Fukushima Pref., TEPCO
			~30 points in offshore area (30 - 90km from coast)	Once a three months	MEXT
			~12 points in ocean (over 90km from coast)	Once a six months on a voyage	JCG, MEXT, Fisheries Agency, JMA
			10 points in Tokyo Bay	Once a several months	MEXT, MLIT, ENV, JCG, Local Government
			5 points in Tokyo Bay	Once a week	MLIT
	Sea Sediment	Sample Analysis	~60 points in coastal sea area (within 30km from coast)	Once a several months	MEXT, Fukushima Pref., TEPCO
			~30 points in offshore area (30 - 90km from coast)	Once a three months	MEXT
			10 points in Tokyo Bay	Once a several months	MEXT, MLIT, ENV, JCG, Local Government
	Public Water	Sample Analysis	~100 points in Fukushima Prefecture and ~10 points in other prefectures	Once a several months	ENV, LNERH
	Natural Park	Sample Analysis	16 points of 4 Prefectures around Fukushima	Once a year	ENV
	Sewage Sludge	Sample Analysis	~ Several hundred points in all over Japan	As needed	Local Government, MLIT
	Waste	Sample Analysis	At waste treatment facilities handling contaminated materials	As needed	ENV, LNERH
	Food	Sample Analysis	All over Japan	As needed	Local Government, Fisheries Agency, MAFF, MHLW
	Water Supply	Sample Analysis	At Water Treatment Plants, Water Intake s or Water Sources	Once a week	Fukushima Pref., MHLW

MEXT: Ministry of Education, Culture, Sports, Science and Technology, MLIT: Ministry of Land, Infrastructure, Transport and Tourism, MAFF: Ministry of Agriculture, Forestry and Fisheries, ENV: Ministry of Environment, JCG: Japan Coast Guard, JMA: Japan Meteorological Agency, LNERH: Local Nuclear Emergency Response Headquarters , TEPCO: Tokyo Electric Power Company, FEPC: The Federation of Electric Power Companies of Japan

Figure 5-2. Monitoring Activities Carried Out in Support of the Comprehensive Plan.

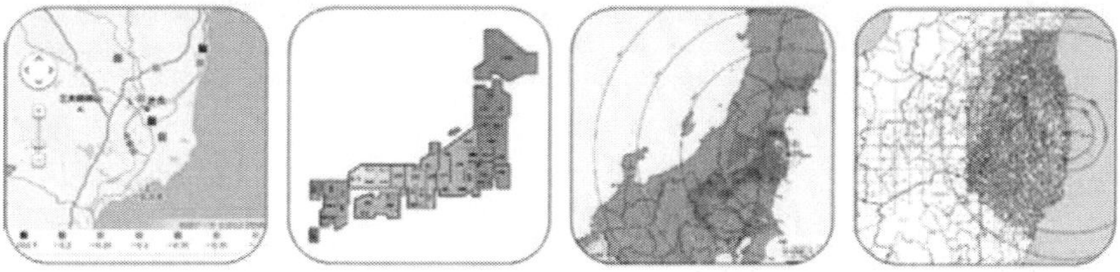

Figure 5-3. Nuclear Regulation Authority has begun to consolidate environmental monitoring information[47].

5.3. Recommendations and Supporting Actions

Recommendation 1

Develop and implement an overall environmental monitoring plan that strengthens the linkage between the purpose/need for data and the data collection and management protocols.

An effective long-term environmental monitoring program needs to be designed with a clear understanding of the decisions that will be made using the data that are collected. For the Fukushima situation there is a very broad range of potential purposes and needs for data collection. These purposes include understanding the nature and extent of contamination, assessing the overall effectiveness of remediation efforts to support re- population decisions, understanding the fate and transport of cesium in the environment to support long-term strategies for protection of human health, etc. Each of these purposes brings its own set of data quality requirements. A well-defined monitoring program will clearly link the data collection requirements to the purpose and need for using the data.

Supporting Actions

- This plan needs to address fundamental QA/QC requirements for all monitoring data. Also the end use or purpose of data collection needs to be clearly specified to ensure that data collection methods, resolution of data, frequency of collection, etc. meet overall needs.
- This plan should address: 1. purpose of the data, 2. predefined data quality objective (what kind of resolution, accuracy, precision, duration, monitoring/sampling interval, etc.), 3. chain of custody (basic information of data, who collected it and where, record of data transfer), 4. quality assurance/quality control (each data set needs to go through predefined QA/QC protocol), 5. reporting (how the data will be reported), 6. Storage (how and where the data will be stored and available).

ESFs suggest the following tools and methodologies that USG uses for site survey and sampling planning; MARSSIM, MARSSIME, DQO, Visual Sample Plan, etc. For example, the USG often uses the Multi-Agency Radiation Survey and Site Investigation Manual (MARSSIM) to planning, implementing, and evaluating environmental and facility radiological surveys to demonstrate compliance with a dose- or risk-based regulation. MARSSIM is developed based on the data quality objective process to plan and make decisions systematically. MARSSIM training is available from the US EPA. Appendix 3 includes a list of relevant resources on this topic.

Recommendation 2

Enhance the data management systems to improve the consistency of data storage methods and accessibility to facilitate visualization and multi-disciplinary data evaluation and analysis.

To date, many separate data storage and access methods are being used. The current efforts by NRA to consolidate Fukushima-related data bases will be very beneficial in supporting a more integrated approach to visualization and analysis of multiple data sets. The current pdf-based approach for capturing and reporting most monitoring Fukushima data sets needs to be quickly replaced with a web-based system for accessing, visualizing and analyzing monitoring data.

Supporting Actions

- Provide implementing guidelines for multiple agencies that are collecting monitoring data.
- Nuclear Regulation Authority should develop a web-based data integration function.

One effective method to alleviate this challenge that is used at the US DOE's Hanford site is through the use of the PHOENIX (PNNL Hanford On-Line Environmental Information Exchange) web-based tool for accessing multiple environmental monitoring data bases. This tool provides technical and regulatory staff with easy access to important data for understanding the changes in environmental conditions and progress of remediation efforts. Appendix 3 provides a link for this resource.

Recommendation 3

Conduct periodic reviews and evaluations of monitoring data to ensure appropriate feedback with other strategic functions including efforts to optimize decontamination strategies, efforts to improve understanding of cesium behavior in the environment, and efforts to optimize the long-term monitoring program.

A continuing theme of this report is that the elements of the overall remediation system are inter-connected. Monitoring data and its evaluation for decision making purposes needs to be routinely reviewed with organizations involved in designing radiation protection strategies, planning remediation strategies, and evaluating cesium transport in the environment. The monitoring coordination efforts are extremely valuable. An analogous effort should be initiated to routinely evaluate monitoring data.

Supporting Actions

- Maintain linkage to radiation protection to ensure that radiation protection strategies are adapted to the most current and relevant conditions.
- Maintain linkage to remediation strategy so that priorities for action can be focused on the most important aspects of remediation.
- Maintain linkage to cesium behavior in the environment so that models can be validated and anomalous monitoring observations can be investigated.

The US DOE has initiated an effort entitled "Scientific Opportunities in Monitoring of Environmental Remediation Sites" (SOMERS). An initial publication provides DOE's vision for advancing monitoring through an integrated systems-based approach. This document identifies in detail scientific and technical challenges and opportunities associated with systems-based monitoring at DOE sites. This continuing effort can support design of more effective long-term monitoring strategies.[48] A link to this information is provided in Appendix 3.

6.0. ASSESSMENT OF REMEDIATION SYSTEM PROGRAM ELEMENTS: CESIUM BEHAVIOR IN THE ENVIRONMENT

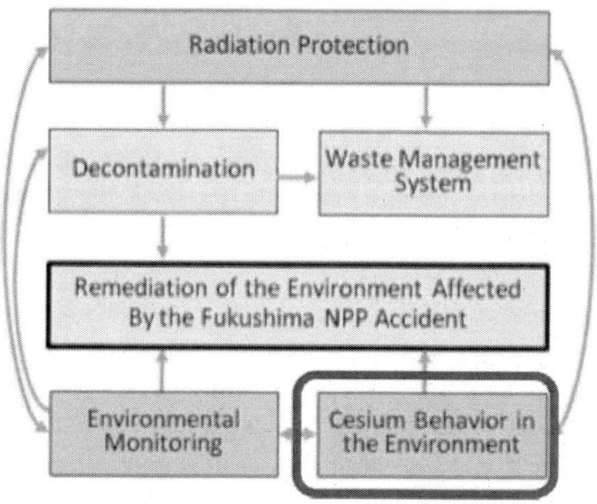

The distribution of radioactive cesium on land will change with time. Cesium can be released from environmental media (e.g., soil) to which it is presently attached physically and/or chemically, and transported by various mechanisms (dissolved, colloidal, aerosol mechanisms). It can be subsequently absorbed into and released from other environmental media, including biota. Prediction and monitoring of its transport and fate is needed to evaluate the potential for accumulation and increased dose and the impact of radiation dose on the indigenous biota.

6.1. Current Situation

Natural attenuation through both radioactive decay and by dispersal is expected to reduce the concentration of cesium at most locations with time. However, accumulation of radioactive cesium at environmental features, such as riverbanks, can occur and cause "hot spots" or high dose rates at those features. Similarly, accumulation into biotic species can also occur and thereby increase the dose to, and potentially impact the health of, the species. For this reason, it is important to understand the mechanisms and kinetics of transport, and identify the accumulation points and the fate of the cesium.

The evaluation of radioactive cesium movement in the environment, and the impact of dose on biota is generally performed through scientific investigations. Analytic models are often used to evaluate the transport, deposition, and the fate of cesium in the environment. Monitoring for cesium at general locations can be used to demonstrate natural attenuation; monitoring for cesium at predicted accumulation points can be used to benchmark the models for transport and accumulation. Measurement of cesium uptake in biota, and the evaluation of the impact of the dose from cesium radiation can be investigated. Characterization of the chemical, physical, and biological relationships to the radioactive cesium distribution pattern in a contaminated land region can be compared to a similar non-contaminated land region to critically evaluate the effects of radioactive cesium contamination on the environment.

The application of the knowledge and insights gained from the scientific investigations can provide the technical bases and guidance for the remediation strategies and decontamination targets. Examples include:

- The applied knowledge, using analytic transport models, that show cesium transport through man-made and/or natural barriers is very low could be used to demonstrate the safety in the design of a waste storage or disposal facility.
- The applied knowledge that cesium is rapidly transported as a dissolved element in water and accumulates in soils at riverbanks could be used to place cesium collection media at these locations.
- The applied knowledge of a lack of cesium uptake from soils into certain crops could be used to enable these crops to be safely grown and harvested.

- The applied knowledge that specific plant species greatly accumulate cesium could be used to accelerate remediation of a land through a sequence of planting and harvesting.
- The knowledge acquired through monitoring and trending of that show no adverse effects of very low-level radiation on biota could be used to justify decontamination targets.

There would be many other practical applications of the knowledge acquired from scientific investigations to support and accelerate remediation of the environment and justify the establishment of decontamination targets in a remediation system.

6.2. Summary of Observations

Scientific investigations have been and are being undertaken to provide the data and establish a knowledge base to understand cesium transport and its fate, including effects on biota. The GOJ has sponsored research and studies on the behavior of cesium in the environment with staff at the National Institute of Environmental Studies (NIES) and the Japan Atomic Energy Agency (JAEA). Research under the main project with NIES, "Behavior of radionuclides in the environment and their effects on living organisms after the Fukushima Daiichi Nuclear Power Plant accident," is in progress. The areas of research are comprehensive and include:

- Multimedia environmental monitoring
- Field monitoring
- Ecosystem impact evaluation
- Long-term human exposure estimation
- Analytic method development (for ultra-sensitive cesium detection)

To date, the NIES has provided important information to date in regards to understanding:

- Initial contamination distribution based on prevailing weather patterns
- Initial hold-up of cesium in the soils[49]

Recent reports and papers of the scientific investigations performed to date are listed in the documentation in Appendix 2.

The GOJ-MOE has planned to sponsor a new institution, the "Center for Environmental Creation" with researchers from JAEA and NIES. This new institution will provide new, dedicated research facilities and research teams for remediation topics and also to provide education services. The GOJ-MOE intends to look to other academic institutions in Japan and U.S. partners to guide and participate in the research activities to be performed at the Center.

The U.S. has had cases of cesium contamination at its production reactor sites that occurred in the 1950s and 1960s. Specifically, contamination of the wetlands and forested regions at the Savannah River Site that came from inadvertent releases during reactor operation. Investigations have been performed to understand the distribution and impact of cesium in the environment with time[50,51,52,53,54,55,56,57,58]. The topics of these investigations included cesium concentration and transport mechanisms for atmospheric, surface water, and ground water; monitoring techniques for cesium in biota; characterization of the distribution of cesium in biota, and remediation technologies.

6.3. Recommendations and Supporting Actions

Recommendation 1

Continue development of cesium fate and transport models to enhance the ability to predict cesium movement and accumulation in the affected environment.

The GOJ should continue to sponsor research and studies on the behavior of cesium in the environment. This work would include further development of environmental models and other scientific investigations to enhance the capabilities to predict cesium transfer from the contaminated areas to other areas, and to evaluate the impact of radioactive cesium on the environment, including the biota. The following basic areas are suggested to improve the prediction capability for cesium transport, verification of model results, and evaluation of effects of cesium on biota:

1. Evaluation and comparison of existing and/or new models for the prediction of cesium transfer through natural and man-made media, and for its fate (accumulation in media and biota)
2. Monitoring and sampling for cesium at selected locations in the contaminated and non-contaminated sites; remediated sites; storage

and disposal sites; and in biota to evaluate: 1) accumulation and 2) natural attenuation; 3) re-contamination; 4) cesium transport; 5) cesium uptake

3. Benchmarking and/or verification of the models using monitoring and sampling data
4. Identification of key parameters that are indicators for impact to humans and other biota in the environment, and characterization of dose/biota effect relationships.

The contaminated areas in Japan cover various ecosystems including forest, marine, urban and rural environments. As an initial step, it is recommended to develop or improve the model for the specific environment. Each model development needs to have the specific purpose for its application. For example, the forest areas that are not decontaminated may release cesium to the nearby residential or business area, which potentially increase the external exposure to the public. A forest environment model would provide insight to cesium transport and fate in that environment. A marine environment model can focus on predicting the accumulation of cesium in sediments and organisms specific to that environment. The model output can be used to estimate any harmful effects to the marine wildlife and food product from marine environment.

Because environmental models typically involve with the multiple expert areas, it is recommended to work closely with other research centers within Japan and also international collaboration. The U.S. EPA and DOE national laboratories are good collaborative partners to exchange expertise and resources. The collaborations can be on specific research proposals to provide either complementary research on a topic or confirmatory research on a finding. Other forums such as workshops, webinars and online trainings, scholar exchange program, and meeting in the international forums, can facilitate the sharing of knowledge and understanding of cesium behavior in the environment.

Supporting Actions

- Identify key parameters and relevant to understanding impacts to human health and the environment.
- Directly link model development to monitoring of the affected environment.
- Calibrate models using observed data
- Assess the selected models using monitoring and sampling data

Recommendation 2

Develop and apply models to evaluate and enhance the effectiveness of decontamination strategies and technologies.

This recommendation would be executed in conjunction with evaluation of advanced decontamination technologies described in Chapter 3 of this report.

The work would involve evaluation and modeling of the physical and chemical binding and release mechanisms for cesium from natural and man-made material systems. This work would also involve the investigation and evaluation of decontamination using biota as the decontamination agent and amendment agents to avoid uptake in biota.

Supporting Actions

- Evaluate mechanisms related to removal of cesium from urban environments.
- Evaluate mechanisms related to remediation of agricultural lands including approaches to block biological uptake of cesium in agricultural products.
- Evaluate mechanisms related to remediation of forest areas include a broad range of phytoremediation approaches.

Recommendation 3

Develop and apply models to inform urgent radiation protection strategies for people living in areas with residual contamination (re-entrainment) and for re-population of evacuated areas.

This recommendation complements the recommendation under Radiation Protection that is to establish a radiation dosimetry program. While a radiation dosimetry program would remove the uncertainty in dose received vis-à-vis models for estimated dose received for members of a population, the dosimetry program may be cost-prohibitive to continue to implement. Improvement to the existing model for dose described in Chapter 2 of this report. The model described in Chapter 2 that is used for decontamination targets can certainly be improved.

Supporting Actions

- Develop population dose/exposure models that are relevant to the specific circumstances, life-styles and exposure pathways that are actually present.
- Evaluate potential recontamination mechanisms due to terrestrial transport of contaminants into areas that have undergone decontamination.

Recommendation 4

Develop and apply models to guide long-term monitoring approaches that will enhance the long-term understanding of cesium (and other contaminants) behavior in the environment.

This recommendation complements the recommendation in Chapter 5. This recommendation calls for a framework to optimize monitoring and sampling methods and strategies specific to cesium in the environment.

Supporting Actions

- Develop and maintain a direct interaction with the design of long-term monitoring efforts. Models should be used to define priorities for environmental sampling and analysis including identification of indicator species (sentinels), etc.

Recommendation 5

Investigate cesium effects on environmental receptors.

This recommendation is for establishment of an information base for dose-health effects records. This would further inform the decision of decontamination targets or dose reduction standards discussed in Chapter 2 of this report.

Supporting Actions

- Develop and maintain a dose - health dataset (that maintains individual privacy) to augment the international body of data used to evaluate health effects from radiation

7.0. Assessment of Remediation System Program Elements: Remediation of the Environment Affected by the Daiichi NPP Accident

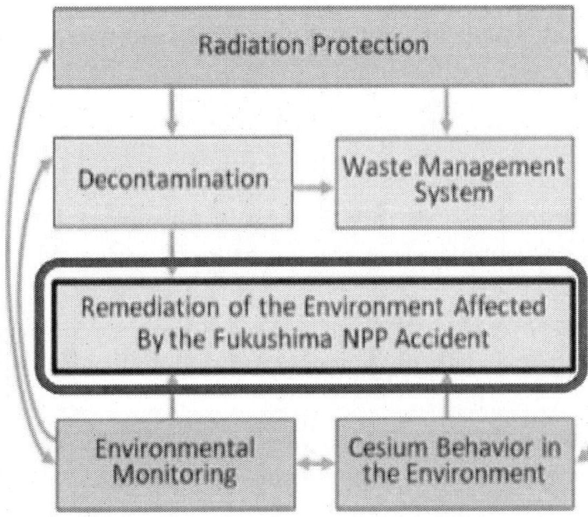

The environmental remediation strategy defines the overall priorities and sequence for applying decontamination resources to specific problems types and locations within the affected regions. This strategy should guide the overall pace of work and should be revisited periodically as decontamination progress is made. This strategy will be especially important as plans are set for remediation of the highest dose regions within the evacuation zone (Special Decontamination Area). The remediation strategy will need to be closely coordinated with plans for re-population and reconstruction of the communities that have been evacuated.

The following sections provide an overview of the current remediation strategy for Fukushima off-site cleanup including the status and progress accomplished to date. Key findings and recommendations are also provided.

7.1. Current Situation

Japan has defined two areas for organizing remediation activities (see Figure 7-1):

- The **Intensive Contamination Survey Area** (ICSA). The ICSA encompasses more than 100 municipalities in 8 prefectures[59]. These are areas with an air dose rate greater than 1 mSv/year (equivalent to 0.23 μSv/hour or greater) and less than 20 mSv/year. These areas were not evacuated following the accident. Decontamination work is implemented by each municipality with technical and financial support provided by the national government.
- The **Special Decontamination Area** (SDA). These are the areas[60] that were evacuated as a result of the nuclear power plant accident and include the area within 20 km from the Daiichi nuclear plant and the areas where the air dose rate exceeds 20 mSv/year. Decontamination in these areas is implemented by the national government.

Over a three year period, approximately 1.3 trillion yen has been allocated by the National government for post- accident decontamination related work.

Status and Progress within the Intensive Contamination Survey Area

The ESF focus has been on the Special Decontamination Area and the efforts of the Ministry of the Environment to carry out remediation within these regions. Remediation within the ICSA is guided by the long-term goal to reduce additional exposure (i.e., dose from radioactivity released by the accident that is in addition to natural background radiation) to 1 mSv/year or less. Near-term goals include the following:

- Reduce estimated annual exposure of the general public **by 50% in 2 years** (by August 2013) by radioactive decay and by decontamination.
- Reduce estimated annual exposure of children by **60% in 2 years** (by August 2013) by thorough decontamination of their living environment.

The ESFs conducted visits to Fukushima City that included national, prefectural and city government organizations involved in decontamination efforts. Fukushima city is the largest population center within Fukushima perfection with a total population of about 290,000 people. There are 19 districts within the city and the air dose rate varies 0.26 μSv/hour (~1.2 mSv/year) to 2.24 μSv/hour (~11.6 mSv/year), as reported for the period of May to July 2012.[61]

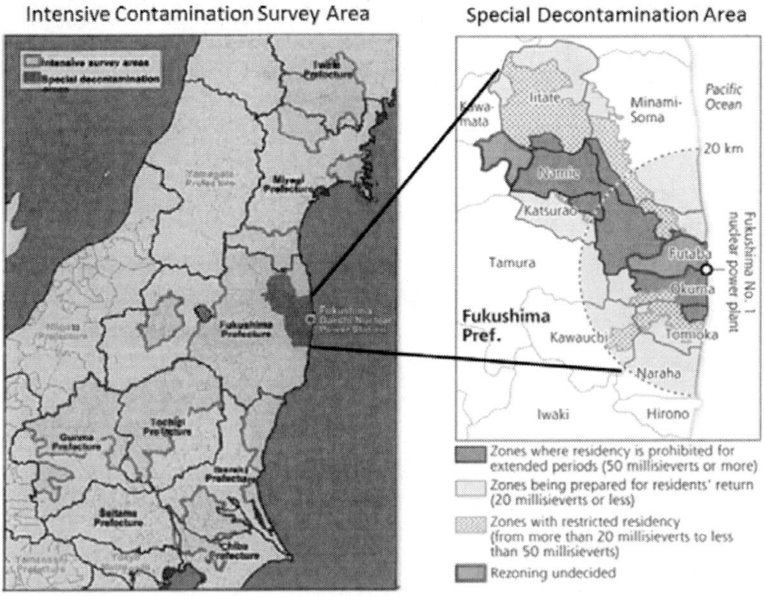

Figure 7-1. Boundaries of Intensive Contamination Survey Area and Special Decontamination Area.

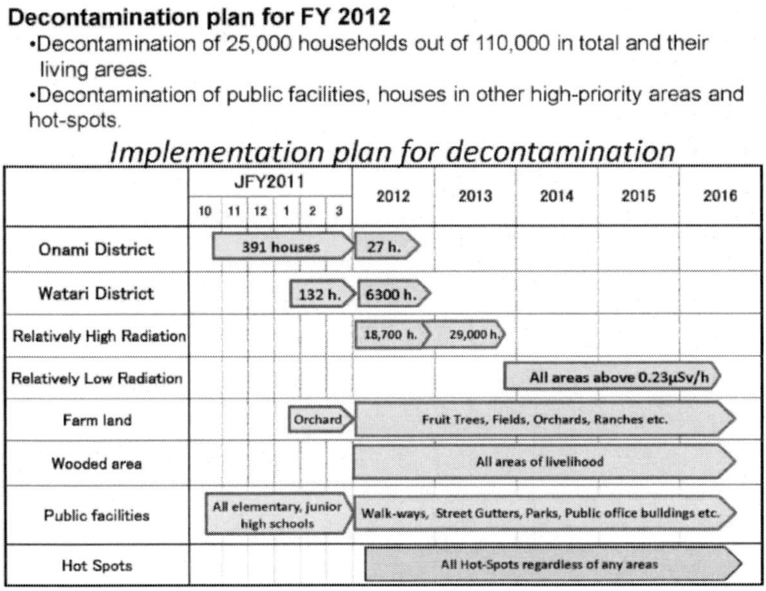

Figure 7-2. FY 2012 Decontamination Plan for Fukushima City.

Fukushima City first prepared a decontamination plan on September 27, 2011 and then revised the plan on May 21, 2012 based on the statutory requirements of the "Act on Special Measures" that took effect in 2012. Priorities for implementing decontamination efforts were assigned to two districts showing the highest air dose rates, Onami and Watari. Priority was also given to areas that would be frequently used by citizens such as roads, schools, and parks.

Current Status of Decontamination in Fukushima City

1 Decontamination 放射線総合対策課

■Houses (As of Feb 1, 2013) (unit : house)

District		(A) Ordered	(B) In process	(C) Completed	(D)＝(B)+(C)	(D)／(A) Progress rate
Onami	1st phase	418	0	418	418	100.0%
	2st phase	52	0	52	52	100.0%
Watari	1st phase	717	0	717	717	100.0%
	2st phase	2,831	336	2,083	2,419	85.4%
	3st phase	2,471	567	228	795	32.2%
Tobu	1st phase	1,027	60	333	393	38.3%
Tatsugoyama	Whole area	557	16	23	39	7.0%
Chuo	1st phase	5,349	88	32	120	2.2%
Iino	Whole area	2,430	19	14	33	1.4%
Matsukawa	1st phase	900	25	1	26	2.9%
Horai	1st phase	2,161	63	0	63	2.9%
Shimizu	1st phase	1,893	0	0	0	0.0%
TOTAL		20,806	1,174	3,901	5,075	24.4%

■Roads (As of Feb. 1, 2013) (unit : Km)

District	(A) Ordered	(B) In Process	(C) Completed	(D)=(B)+(C)	(D)／(A) Progress
Onami (cityroad)	40.0	0.0	40.0	40.0	100.0%
Onami (farmroad)	22.2	6.3	15.9	22.2	100.0%

■Forest (Within Living area) (As of Feb. 1, 2013) (unit : ㎡)

District	(A) Ordered area	(B) Completed	(B)／(A)
Onami	424,970	470	0%
Watari	71,353	42,567	60%

■Farmland (As of Feb. 1, 2013) (unit : ha)

Type of Farmland	(A) Ordered area	(B) Completed	(B)／(A)
Orchards	2,106	2,106	100.0%
Meadows	65	0	0.0%
Puddies	2,397	2,339	97.6%
Fields	1,184	676	57.1%

Figure 7-3. Status of Fukushima City Remediation in Selected Towns as of February 2013.

Objectives for Fukushima City decontamination further refined the national objectives as follows:

- Reduce the air dose rate in citizen's living environment to less than 1 μSv/hour in two years
- Reduce the air dose rate by 60% for areas with less than 1 μSv/hour in two years
- Reduce citizen's estimated exposure to less than 1 mSv/year (0.23 μSv/hour) in the future.

Figure 7-2 shows the implementation Plan for FY 2012 decontamination efforts in Fukushima City. Figure 7-3 shows the status of decontamination efforts as of February 2013 in Fukushima City. Current statistics on degree of completion within Fukushima City are available online at http://josen.env.go.jp/zone/details/fukushima_fukushima.html (Japanese only).

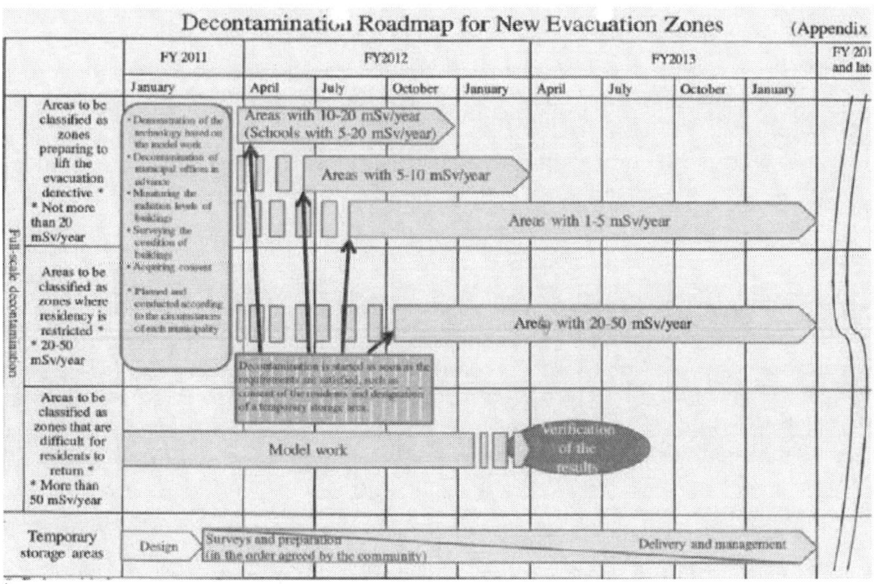

* Each municipal government specifies the concrete decontamination preocedures.

* Technical knowledge obtained through the model work (Cabinet Office and the Ministry of the Environment) is applied to decontamination at the appropriate time.

Figure 7-4. Short-Term Decontamination Roadmap for Special Decontamination Areas.

Status and Progress within the Special Decontamination Area

Figure 7-4 presents the roadmap for decontamination of the Special Decontamination Area. The three portions of this figure correspond to the green (<20 mSv/year), yellow (20 – 50 mSv/year) and red (>50 mSv/year) portions of the Figure 7-1. The basic principles for implementing work in these areas were set by the "Act on Special Measures." For those areas with the dose rate less than 50 mSv/year, the intent is for the national government to develop decontamination plans and "implement decontamination measures for buildings such as houses, business offices and public facilities, roads, farmland, and forests around living areas … by the end of March 2014."

Progress of work in the Special Decontamination Area

	Advance Decontamination (base facilities, etc.)	Identification of owners of houses, etc.	Decontamination plan	Full scale Decontamination Works	Securing temporary storage sites
Tamura city	✓	✓	✓ (Apr 13)	in operation (July 25)	✓
Naraha town	✓	✓	✓ (Apr 13)	in operation (Sept. 6)	✓
Kawauchi village	✓	✓	✓ (Apr 13)	in operation (Sept. 4)	✓
Iitate village	✓	✓	✓ (May 24)	in operation (Sept. 25)	✓ (partly secured)
Katsurao village	✓	✓	✓ (Sep 28)	in preparation (Oct. 12)	✓ (partly secured)
Kawamata town	✓	✓	✓ (Aug 10)	In preparation (Nov. 1)	✓ (partly secured)
Minami-Soma city	✓	✓	✓ (Apr 18)		local coordination process
Namie town	✓	✓	✓ (Nov 21)		local coordination process
Okuma town	✓	✓	✓ (Dec 28)		local coordination process
Tomioka town	✓	✓	local coordination process		local coordination process
Futaba town					

* Decontamination works in a municipality are to be implemented on the premises of *formulation of the decontamination implementation plan and securing of temporary storage sites.*

Figure 7-5. Status of Efforts to Conduct Full-Scale Decontamination work in 11 Municipalities within the Special Decontamination Area.

To prepare for this work, the national government, through Japan Atomic Energy Agency (JAEA), conducted a series of model projects (see Decontamination Methods section) to demonstrate the viability of decontamination approaches for the areas with dose rates below 50 mSv/year.

These model projects provided the basis for developing decontamination plans for these areas. Figure 7-5 summarizes the status of work in the SDA as of December 2012. Decontamination plans have been produced and accepted by 9 of the 11 communities that comprise the SDA but full scale decontamination work is underway in only four of the 11 municipalities. One difficulty has been selection of sites for temporary storage of decontamination waste. Unlike the ICSA, full scale decontamination work cannot begin in the SDA until temporary storage sites have been secured. Moving forward with decontamination in these areas has been challenging because of the need for close coordination with local communities whose citizens have been evacuated and are dispersed about the country. Also, gaining community consensus on locations for temporary storage has not been easy.

Municipality	**Percent of Planned Decontamination Work Completed**			
	Residential Land	**Farmland**	**Forest**	**Roads**
Futaba City (02/28/2013)	99%	100%	85%	99%
Naraha Town (02/27/2013)	24%	13%	43%	0%
Kawauchi Village (02/25/2013)	98%	Planned for FY13	28%	30%
Iitate Village (02/27/2013)	1%	0%	4%	0%

Source: http://josen.env.go.jp/area/index.html

Figure 7-6. Status of model decontamination project in high dose area.

Of the four municipalities with full-scale decontamination work underway, only work in Tamura City is nearing completion (http://josen.env.go.jp/area/details/tamura.html). Forest land refers to a narrow band of 20 meters surrounding living areas.

The final portion of decontamination in work in the Special Decontamination Area is for the high dose areas (>50 mSv/year). Figure 7-6 shows that model decontamination projects in the high dose areas were to have been conducted during FY 2012. But, the procurement to initiate these projects has been delayed and the work is now planned to start in the summer of 2013. Following completion of the model projects in the high dose areas (some time during FY13), MOE will determine the preferred approach and timing for conducting decontamination work in these areas.

Japan is rapidly approaching important milestones for completion of decontamination efforts including dose reduction targets by August 2013 and completion of decontamination work in the Special Decontamination Areas with dose rates below 50 mSv/year by March 2014. Japan has developed substantial capacity for performing decontamination work in a very short span of time. Much work remains in both the ICSA and in the SDA. A clear set of priorities will be needed to focus remediation resources to be applied in the municipal-led decontamination efforts, in the low dose SDA, and in the high dose SDA. For the evacuated areas, the pressures and demands to complete work so as to enable re-population of these areas continue to grow.

7.2. Summary of Observations

Japan has made significant progress in remediation efforts in many of the communities with the ICSA. Work has been slow to begin in most portions of the SDA but is expected to move at a much faster pace during FY2013. Key observations include:

- There is a substantial and growing base of information and experience on decontamination effectiveness and costs resulting from the municipal-led efforts in the Intensive Contamination Survey Areas, from MOE-led efforts in low dose Special Decontamination Area, and from prior model project work in the SDA.
- There are many existing challenges for currently planned decontamination work (e.g., lack of a defined or established waste management system for decontamination wastes such as temporary storage). Also, there are significant budget allocation issues related to expansion of decontamination work. Another potential constraint to expanding decontamination efforts is the lack of trained workers.
- The current March 2014 target date for completion of decontamination work in the SDA (other than the higher-dose areas) does not appear to be feasible based on the current pace of work to date and the likely availability of a suitably trained work force.
- There is considerable uncertainty surrounding the viability of conducting decontamination work in the higher dose portions of the SDA. With dose rates being much higher than have been decontaminated to date, the cost and time required to successfully decontaminate these areas will need to be carefully evaluated.
- There does not appear to be a clearly defined process for evaluating options for full-scale decontamination or for making the national decision of whether and how to proceed for the high dose areas.

7.3. Recommendations and Supporting Actions

Recommendation 1

Conduct a systematic review of the decontamination work that has been completed to date (cost, effectiveness, waste generation, etc.) to provide the information base for extrapolating to implementation of remaining decontamination work.

There is a substantial base of decontamination experience that now exists. This information needs to be collected and systematically evaluated to identify variations in efficiency and effectiveness of the approaches applied across the municipalities of Fukushima prefecture. This information would provide a clear base of experience and information for informing national, prefectural and local decisions about remediation strategies and "best practices" for decontamination. Analysis of this information would provide a basis for making realistic estimates of the remediation resources that are required for those areas that have not yet been initiated.

Supporting Actions

- Summarize the resources required to complete remediation efforts (cost, time, workers) and the impacts of the work (dose reduction, waste generation, etc.) and to be summarized on a city-by-city and prefectural basis.
- Use this set of information to evaluate the relative effectiveness and efficiency of remediation efforts carried out by local communities and MOE-led areas to gain insights into the "best practices" that should be pursued in subsequent work.

Recommendation 2

Develop the baseline definition of the total set of decontamination work that needs to be completed.

Large-scale remediation efforts are aided by having a well-defined baseline of all of the work that needs to be accomplished. MOE is beginning to track and report progress on its web site (www.env.go.jp/en). These reports summarize the percent of units of a particular type of target (houses, schools, farmland, etc.) that have been completed for the communities in Fukushima prefecture. This is an important step but a baseline that is useful for estimating future resource requirements should also contain characteristics of those areas that affect the magnitude of decontamination required, e.g., current dose rates, surface characteristics, etc.

Supporting Actions

- Quickly develop a comprehensive approach for collecting and recording decontamination work progress, dose levels achieved, completion status, and waste generation and accumulation. There needs to be a national system to collect decontamination work results, progress, and remaining work.
- Prepare a routine "progress report" that combines remediation results from all municipal, prefectural and national efforts. Should include cost to date, dose reduction achieved, decontamination waste generated, etc.

Recommendation 3

Develop and maintain an overall remediation strategy complete with life cycle cost estimates, resource allocation strategies (e.g., manpower, etc.), and analysis of critical strategic alternatives.

The primary recommendation from this section recognizes that the large-scale remediation of the contaminated areas will require long-term sustained effort. And, this effort will be conducted in several different types of areas, i.e., municipal led remediation in populated areas and MOE-led remediation in evacuated areas with dose rates below 20 mSv/year, between 20-50 mSv/year, and above 50 mSv/year. The broad scope of remediation that will be required requires careful consideration of the best sequence for conducting remediation and the best allocation of limited resources. There is a strong desire to very quickly achieve decontamination goals in the evacuated areas with dose rates below 50 mSv/year. The next element of the overall remediation strategy that needs to be defined is for the evacuated areas with dose rates greater than 50 mSv/year. A set of alternative approaches for conducting remediation efforts in these high dose areas needs to be defined. Alternatives should include a range of options including a "No Action" alternative, a sequential strategy that tackles the lowest dose and highest value areas first, and a full-scale decontamination strategy. By systematically evaluating these options, an informed national dialogue would be supported regarding the best use of limited resources.

Supporting Actions

- Conduct a “feasibility study” for defining and evaluating options for proceeding with full-scale decontamination in high dose areas. This should be conducted in parallel to the model demonstration projects (6 months of effort) that will be conducted in the high dose areas.
- Use this strategy to guide resource allocation for the Intensive Contamination Survey Area (white zone) and the Special Decontamination Area (green, yellow and red zones).
- Develop an end state definition for determining completion of remediation considering the potential diminishing returns and best use of remediation resources. This definition should be linked to the long term environmental monitoring information.

Figure 7-7 depicts the three primary recommendations from the ESF’s review and illustrates aspects of implementing each of the recommendations based on best practices from the ESF’s remediation experience in the US.

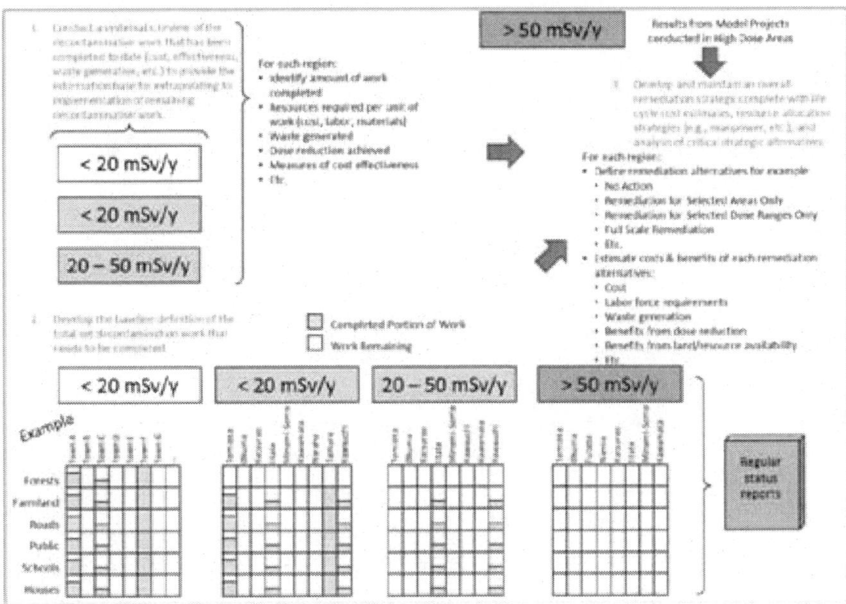

Figure 7-7. Key Recommendations and Elements of an Approach for Maintain an Overall Remediation Strategy.

8.0. ASSESSMENT OF REMEDIATION SYSTEM PROGRAM ELEMENTS: CROSS-CUTTING CONSIDERATIONS

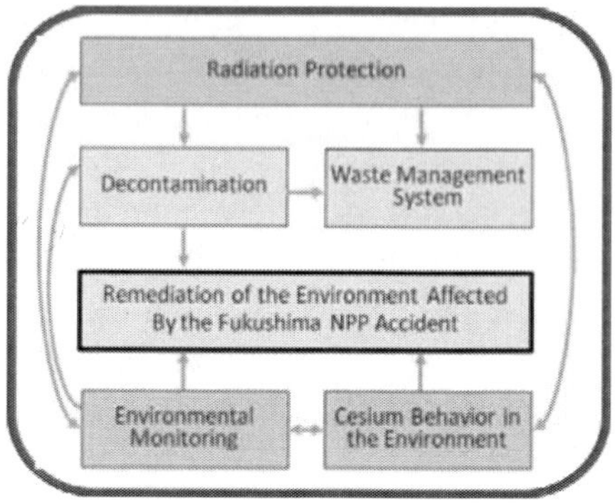

Some aspects of the ESF's observations are important factors in several of the elements of the Fukushima remediation system. For example, effective public engagement in remediation decision processes appears to be lacking in many communities and in several aspects of remediation, e.g., waste storage site selection, use of incineration for volume reduction, and decision making for re-population of evacuated regions.

8.1. Current Situation

Public participation is required to effectively implement many aspects of Fukushima remediation. These aspects include acceptance or consideration of:

- Temporary storage facility sites
- Interim storage facility sites
- Incineration of specified and designated waste (other than decontamination waste)
- Community-specific decontamination plans
- Acquisition of land for building an interim storage facility
- Re-population and reconstruction plans for evacuated communities.

The ESF review did not directly interact with public, community and stakeholder representatives. These observations are the result of discussions with national, prefectural, and community government officials and with decontamination contractor staff.

Site selection of Temporary Storage Facilities (TSFs) for the Intensive Contamination Survey Area (ICSA) is not required for decontamination work to begin or continue. It is clear from statistics provided by Fukushima prefecture that a large portion of the decontamination waste still resides in on-site storage. The national and prefectural government experts have provided information on facility designs. But, residents are reluctant to accept storage areas in their neighborhoods. The phenomenon of "not in my back yard" (NIMBY) is certainly prevalent. But, another factor cited as contributing to this concern is the lack of trust in the government's commitment to begin to remove the decontamination waste to an Interim Storage Facility (ISF) within 3 years. There is a perception that TSFs could be needed for much longer than 3 years.

Table 8-1. Aspects of Fukushima Remediation Effort Requiring Effective Public Involvement

Remediation System Element	Public Involvement Issue
Radiation Protection	• Development of radiation protection strategies for populated areas • Development of re-population guidelines for currently evacuated areas • Development of radiation protection strategies for areas to be re-populated
Waste Management	• Selection of sites for temporary storage facilities in ICSA. • Selection of sites for temporary storage facilities in SDA. • Selection of sites for interim storage facilities • Incineration for treatment of radioactive waste • Reuse or recycle of decontaminated materials
Remediation Strategy	• Development of priorities for remediation of evacuation areas including high dose areas • Definition of remediation and reconstruction efforts for evacuated areas
Environmental Monitoring	• Understanding variations in dose rates and approaches for managing radiation exposure

By not having operating TSFs, community-led decontamination work generates waste that is simply stored at the sites where it is generated, including schools, private homes, parks, agricultural fields, etc. Citizens clearly want decontamination work to be carried out but many are unwilling to accept TSFs in their communities that would consolidate and contain the large volumes of decontamination waste that are being generated. Some communities have been more effective than others in gaining acceptance for TSFs to facility decontamination progress and to serve the overall good of the community.

Table 8-1 provides a summary of the aspects of the Fukushima remediation effort that require effective public involvement.

8.2. Summary of Observations

The principal observations pertaining to public involvement practices are the following:

- Effective public involvement mechanisms (framework, process, and programs) to support large-scale remediation activities have not been widely developed and implemented.
- There is significant variation among municipalities in how effective they are in involving their citizens in remediation decisions.
- There are significant disconnects between public expectations and the reality of decontamination work, e.g., the pace of decontamination and restoration of evacuated areas.
- GOJ agencies do not have experience in establishing effective public involvement institutions to support remediation.
- Effective public participation in broad remediation decision making seems to be aggravated by unresolved compensation issues related the Daiichi nuclear plant accident and the resulting loss of homes and livelihoods for those residents that were evacuated.
- GOJ has engaged numerous advisory groups on matters associated with radiation protection, decontamination methods, cesium behavior and other technical and scientific topics. But, these efforts, while essential, do not substitute satisfy the need for effective public involvement.

- There are multiple communities, very diverse interests and value sets, and a variety of decisions and topics that that make designing effective public involvement strategies very difficult.
- The roles and responsibilities for remediation decision making and oversight of remediation efforts in Japan are significantly different from the roles typical of US DOE cleanup sites and US EPA Superfund cleanup sites. Specifically, the Japan Ministry of the Environment sets standards for decontamination (cleanup goals), develops guidelines for carrying out decontamination work, directs the implementation of the work (directly for the SDA and indirectly through community governments in the ICSA), and verifies that decontamination work has been completed as required. In US DOE experience these functions are typically carried out by DOE, the US EPA, and a state regulatory agency. The US model for site-specific advisory boards, or citizen advisory boards, provides advice to the multiple agencies responsible for remediation work and oversight.

8.3. Recommendations and Supporting Actions

Recommendation 1

There is an immediate need to develop more effective processes for public involvement in remediation system decisions (e.g., site selection for treatment and storage facilities, re-population strategies for evacuated areas).

Fukushima off-site remediation requires a very diverse set of decisions for which public acceptance is required. But, there are many diverse community/public interests and values that complicate consensus building. An important step in moving from the current situation to a more effective setting for public involvement would be for GOJ to charter an independent entity to evaluate practices in communities that have been successful in reaching a consensus on site selection for Temporary Storage Facility and then to work to transfer those practices to communities that have not yet been successful. Having community residents recognize and participate in decisions that move overall remediation efforts forward can create the momentum for transferring these practices to longer-term remediation, reconstruction and re-population decisions.

Supporting Actions

- Public involvement practices and consensus building effectiveness have varied significantly across the communities undergoing decontamination. GOJ should review the variations in these practices and identify the factors that lead to success and that could be applied more broadly within the affected areas.
- For the communities within the evacuation area (Special Decontamination Area) Consider instituting a community/stakeholder involvement process (potentially formally chartered working groups similar to US citizen advisory boards at large cleanup sites) to provide advice on all aspects of remediation. Advice should support:
    - Community-specific input to radiation protection guidelines for re-population of evacuated areas
    - Priorities for implementing decontamination efforts
    - Waste management strategies including site selection for treatment and storage facilities
    - Other aspects of community reconstruction and restoration.

US environmental remediation efforts have extensive experience with various methods for engaging public and stakeholder groups. The public involvement is a process to integrate the knowledge and opinions of others into its decision making. Effective public involvement can improve the content of the government's decisions and the deliberative process. Further public involvement ensures democracy and civic engagement, and builds public trust in government.

US DOE has a charter for Environmental Management Site-Specific Advisory Boards (http://cab.srs.gov/library/charter.pdf). These boards (also known as citizen advisory boards) support remediation efforts at many of DOE's environmental remediation sites including the Savannah River Site in South Carolina and the Hanford Site in Washington state. These boards provide advice and recommendations for site-specific issues such as:

- Cleanup standards and environment restoration
- Stabilization and disposition of non-stockpile nuclear materials
- Excess facilities
- Future land use and long-term stewardship
- Risk assessment and management
- Cleanup science and technology activities.

US EPA's public involvement policy is to improve the effectiveness of EPA's mission by ensuring well-informed decisions, and encouraging innovative methods for involving the public. The public involvement policy in US EPA has the following basic steps.[62] These steps cover all types of public involvement. The goal, actions, and methods of each step can be found in the policy document.4

Step 1: plan and budget for public involvement activities;
Step 2: identify the interested and affected public;
Step 3: consider providing technical or financial assistance to the public to facilitate involvement;
Step 4: provide information and outreach to the public;
Step 5: conduct public consultation and involvement activities;
Step 6: review and use input, and provide feedback to the public; and
Step 7: evaluate public involvement activities.

The goal of the final step is to evaluate the effectiveness of this Policy and of public involvement activities. To implement this action, agency officials should evaluate and measure, on a continuing basis, both the effectiveness of the Policy to improve public involvement in regulatory and non-regulatory processes and the effectiveness of public involvement activities. Agency officials should routinely use surveys, interviews, focus groups and tools to evaluate whether public involvement practices are performed appropriately and have the intended effects (subject to the Paperwork Reduction Act). Agency officials also should conduct periodic broad- based Agency-wide evaluations to determine whether implementing this Policy improves the quality of public involvement and environmental decisions.

The case studies using these steps can be found in the following document published by US EPA: Better Decisions through Consultation and Collaboration, Conflict Prevention and Resolution Center.[63,64] The relevant example to the Fukushima NPP cleanup effort is an EPA Superfund site process. EPA regulations require community involvement throughout the process of the EPA Superfund site cleanup. For example, the community involves in the process by following activities: providing any information about the site to US EPA, participating in the public meetings or EPA events (ask questions, and provide comments on plans for clean up), informing EPA about how the community wants the site to be used in the future, visiting the site to observe cleanup activities, etc. In addition to these required activities, US EPA promotes the public involvement through the following programs:

Technical assistance grants (TAGs) provide money for activities that help the community participate in decision making at eligible EPA Superfund sites.[65] A *community advisory group* (CAG) is made up of members of the community for information exchange among the local community and EPA, the state regulatory agency and other Federal agencies of the EPA Superfund site.[66] *Technical assistance services for communities* (TASC) provide non-advocacy technical assistance services at no cost to communities to empower them to substantively participate in addressing environmental issues and actions which impact their community.[67] *Superfund job training initiative* (SuperJTI) is a job readiness program that provides training and employment opportunities for people living in communities affected by EPA Superfund sites.[68] The regional public liaison (RPL) program provides help people with issues or concerns about EPA Superfund site cleanups. Each regional office in EPA has a Public Liaison.

Superfund Community Involvement Toolkit ("CI toolkit") provides EPA Superfund Regional site teams, community involvement staff, and others with a practical easy-to-use aid for designing and enhancing community involvement activities.[69] The community involvement information in the EPA Superfund site can be found in the official US EPA website: http://www.epa.gov/superfund/community/index.htm.

One of the first US DOE citizen advisory boards was established at the Hanford site. In 1992 the US EPA, US DOE and the Washington (state) Department of Ecology (Ecology) initiated an effort to form an advisory board to address Hanford cleanup issues. An independent environmental mediation and conflict management organization (The Keystone Center) was asked to explore the potential to form an advisory board. This group issued a convening report in October 1993, http://www.hanford.gov/files.cfm/Convening_Report.pdf. The methods used to identify the appropriate interest groups and to identify the basic expectations for a successful board are also applicable to the current Fukushima off-site remediation efforts. Among the interests to be represented were the following:

- Local citizen and governmental interests
- Local business interests
- Local environmental interests
- Labor/work force interests
- Regional environmental, citizen and other public interests
- Regional business interests
- Tribe with ceded lands on or adjacent to Hanford

- The State of Oregon; and
- The general public.

The convening report identified the anticipated benefits of an advisory board for DOE, EPA and Ecology including:

- Be a well-informed group of local, regional, and tribal representatives who are focused on problem solving and proving input on key policy issues;
- Improve open communications between and among board members, the sponsoring agencies, and the public;
- Provide broader, more robust definitions of problems, priorities and alternatives;
- Help the agencies reach key decisions and set priorities in an era of tight budget constraints;
- Provide a forum in which the agencies are publicly accountable for progress on Hanford cleanup and compliance with all applicable state and federal laws'
- Advise agencies on how to coordinate and carry out these activities in ways that maximize public involvement opportunities and minimize unnecessary duplication and conflicts in scheduling and contribute to agency decisions that better reflect the principles and values of all of the diverse Hanford interests.

Equally important were the agencies' collective commitments to the advisory board. The agencies state that they will:

- Not attempt to control the Board or its agenda;
- Treat Board members with candor and respect;
- Listen to and try to understand Board members' views;
- Honor, respond and give serious consideration to the views, recommendations and advice of the Board in agency policy development, decisions and actions'
- Provide sufficient notice to the Board regarding emerging issues and imminent policy decisions in time for the Board to make a choice about whether it wishes to provide recommendations and advice on the decision and/or the manner in which the broader public should be involved in the decision;

- Work with the Board to provide funds for independent technical assistance, staff and other administrative support, facilitators, and access to information and agency personnel.

While US cleanup institutions have extensive experience with various public involvement mechanisms, the ESFs recognize that these approaches would not translate exactly into Japanese society. But, these practices could be adapted for use in Japan and merged with successful models of effective participation that have emerged following the start of Fukushima off-site remediation efforts. Also, the concept of a citizen advisory board, if adapted to Fukushima remediation, could address specific remediation issues such as cleanup standards, remediation priorities and methods, and storage site selection. But this mechanism could also address related issues such as reconstruction of vital infrastructure, restoration of local economic and social well-being, and strategies for re-population of evacuated areas.

Additional resources are identified in Appendix 3.

Appendix 1. ESF Biographical Statements

Sang Don Lee is a Research Environmental Scientist for the United States Environmental Protection Agency's Office of Research and Development. He has 9 years of experience at the EPA in decontamination and consequence management. His expertise includes material engineering, aerosol science, and environmental science, and much of his research at the EPA has focused on the fate and transport of radionuclides in the urban environment. Dr. Lee received his Ph.D. in Environmental Sciences and Engineering in 2004 from the University of North Carolina at Chapel Hill after earning his Master's degree in Environmental Engineering from Korea University in 1998.

Robert (Bob) Sindelar is a Senior Advisory Engineer in Materials Science and Technology at the Savannah River National Laboratory. He has 28 years of experience at SRNL in research, development, and deployment activities in demonstration of the structural integrity of aging nuclear materials systems and in evaluation of the fuel and structural materials to enable the safe transportation, storage, treatment, and disposal of spent nuclear fuel. Dr. Sindelar is a leading international expert in nuclear science, including spent nuclear fuel management, water decontamination, aging effects and aging management of structures, and materials in nuclear systems. He leads the technology programs to provide for the safe management of research reactor fuel in wet and dry storage systems at the Savannah River Site and worldwide through consultancies with the International Atomic Energy Agency. He also supports the extended dry storage and transportation of commercial reactor fuel for the DOE Office of Nuclear Energy. Dr. Sindelar received his Ph.D. in Nuclear Engineering in 1985 from the University of Wisconsin.

Mark Triplett is a Senior Advisor at the Pacific Northwest National Laboratory. He has more than 30 years of experience at PNNL in waste management systems analysis, decision analysis for environmental remediation, and risk communication. Mr. Triplett has worked on many aspects of the Hanford site cleanup, including soil, groundwater and tank waste cleanup, and has expertise in planning and prioritizing cleanup activities as well as in communicating cleanup issues to stakeholder groups. Mr. Triplett presently is supporting the DOE Richland Operations Office and Office of River Protection by developing overall cleanup strategies and by integrating efforts to remediate contaminated soil and groundwater at the Hanford site. Mr. Triplett received his M.S. in Engineering Science in 1975 from Purdue University.

APPENDIX 2. BIBLIOGRAPHY OF DOCUMENTATION REVIEWED

[1] Hideki Kawamura, *Obayashi Corporation, Obayashi Activities on Environmental Remediation and Interim Storage Project*, 3/6/2013.

[2] Ministry of the Environment, *Local Contamination by Radioactive Materials to Guidelines for Dealing With*, 3/2012.

[3] Ministry of the Environment, *Decontamination Guidelines*, 1st ed.. 12/2011 (Tentative Translation), mm/dd/yyyy

[4] Ministry of the Environment, *Removing Dye Relationships Q&A*, 14th ed., 2/2013

[5] Ministry of the Environment, *Japan, Japan: Standard for Remediation*, 2/5/2013

[6] Ministry of the Environment,
除染関係ガイドライン,
平成23年12月 第1版

[7] Ministry of the Environment,
汚染状況重点調査地域内における
環境の汚染状況の
調査測定方法に係るガイドライン
平成23年12月 第1版

[8] Ministry of the Environment,
除染等の措置に係る
ガイドライン
平成23年12月 第1版

[9] Ministry of the Environment,
除去土壌の収集・運搬に係る
ガイドライン
平成23年12月 第1版

[10] Ministry of the Environment,
除去土壌の保管に係る
ガイドライン
平成23年12月 第1版

[11] Risk Management Office, Fukushima City, "*Decontamination in Fukushima City*," 3/2013.

[12] Risk Management Office, *Fukushima City, Fukushima Dosimeter Measurements*, Exhibit 2, mm/dd/yyyy

[13] Fukushima municipal measures crisis management room Radiology General measures Division, Hanzawa, *Collaborates with District Town Council Pertaining to the Implementation of Decontamination (Collaboration)*, mm/dd/yyyy

[14] Risk Management Office, Fukushima City, *How to Advance Housing Decontamination*, mm/dd/yyyy

[15] Risk Management Office, Fukushima City, *Decontamination in Fukushima City*, 3/2013

[16] Risk Management Office, Fukushima City, *Nuclear Disaster Measures (Other Than Decontamination)*, mm/dd/yyyy

[17] Environment Division Fukushima Prefecture except decontamination measures, *Except Decontamination Measures in Fukushima Prefecture*, 3/7/2013

[18] Risk Management Office, Fukushima City, *How to Advance Housing Decontamination*, mm/dd/yyyy

[19] Fukushima municipal measures crisis management room Radiology General measures Division, Hanzawa, *Collaborates with District Town Council pertaining to the Implementation of Decontamination (Collaboration),* mm/dd/yyyy

[20] Yoichi Ito, Director, Japan Atomic Energy Agency,: *(2) Report of the Results of the Decontamination Model Project: Overview of the Results of the Decontamination Model Projects - Role and Overview of the Model Projects,* 3/26/2012

[21] Keiichi Kawase, Japan Atomic Energy Agency,: *(2) Report of the Results of the Decontamination Model Project: Overview of the Results of the Decontamination Model Project – Overview of the Results of Group A,* 3/26/2012

[22] Mitsugu Kato, Japan Atomic Energy Agency,: *(2) Report of the Results of the Decontamination Model Project*: *Overview of the Results of the Decontamination Model Project – Overview of the Results of Group B,* 3/26/2012

[23] Kazuki Iijima, Japan Atomic Energy Agency,: *(2) Report of the Results of the Decontamination Model Project: Overview of the Results of the Decontamination Model Project – Overview of the Results of Group C,* 3/26/2012

[24] Shinji Kihara, Japan Atomic Energy Agency,: *(2) Report of the Results of the Decontamination Model Project: Overview of the Results of the Decontamination Model Projects,* mm/dd/yyyy

[25] Shinji Kihara Japan Atomic Energy Agency, - *Overview of the Results of Decontamination Demonstration Tests Conducted in Date City and Minami Soma City,* 3/26/2012

[26] Masayuki Ito, Japan Atomic Energy Agency: (2) Report of the Results of the Decontamination Model *Project*: Analysis and Evaluation of the Results of the Decontamination Model Projects - Decontamination *Technologies,* 3/26/2012

[27] Shinichi Nakayama, Japan Atomic Energy Agency: (2) Report of the Results of the Decontamination Model Project: Analysis and Evaluation of the Results of the Decontamination Model Projects - Decontamination *Wastes (Removed Objects) and Their Temporary Storage,* 3/26/2012

[28] Tomohiro Asano, Japan Atomic Energy Agency: (2) Report of the Results of the Decontamination Model Project: Analysis and Evaluation of the Results of the Decontamination Model Projects - Radiation Control, 3/26/2012

[29] Yoshitake Shiratori & Akihiro Tagawa, Japan Atomic Energy Agency: *(2) Report of the Results of the Decontamination Model Project: FY 2011 "Decontamination Technology Demonstration Test Project",* 3/26/2012

[30] Taiheiyo Cement Corporation, Contractor, Japan Atomic Energy Agency: *Development of Thermal Cesium Removal Technology for Contaminated Soil,* mm/dd/yyyy

[31] Fukushima Environmental Safety Center Headquarters of Fukushima Partnership Operations, Japan Atomic Energy Agency: *JAEA's Activities Toward Environmental Remediation in Fukushima,* 2/13/2013

[32] Japan Atomic Energy Agency, *JAEA Dose Reduction Evaluation,* mm/dd/yyyy

[33] Japan Atomic Energy Agency, *JAEA Dose Reduction Evaluation,* mm/dd/yyyy

[34] Japan Atomic Energy Agency, *JAEA's Activities for Restoration from TEPCO's Fukushima NPP Accident,* 2/8/2013

[35] Japan Atomic Energy Agency, *JAEA's Activities for Restoration from TEPCO's Fukushima NPP Accident,* 2/8/2013

[36] Japan Atomic Energy Agency, *JAEA's Activities for Restoration from TEPCO's Fukushima NPP Accident,* 2/8/2013

[37] Fukushima Environmental Safety Center Headquarters of Fukushima Partnership Operations, Japan Atomic Energy Agency, *JAEA's Activities toward Environmental Remediation in Fukushima,* 2/13/2013

[38] Japan Atomic Energy Research Development Organization, Fukushima Technology Development Division, *Study on Ultra-high Pressure ⍰the Dyeing Technology,* 2/18/2013

[39] Japan Atomic Energy Research Development Organization, Fukushima Technology Development Division, *Study on Ultra-High Pressure ⍰the Dyeing Technology,* 2/18/2013

File: MOE Decon Technologies and Performance/MOE decontamination effectiveness from MOE Fuku office

[40] Ministry of the Environment,

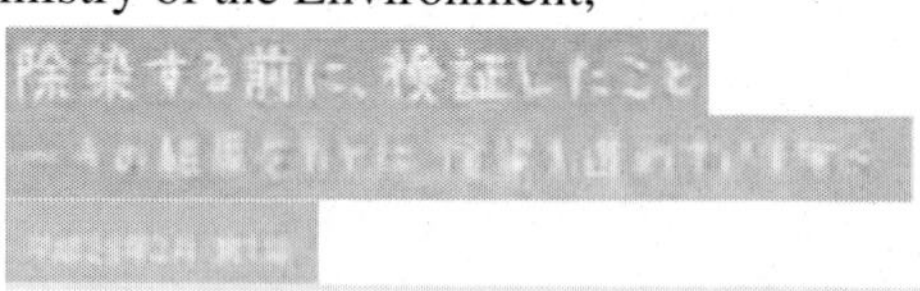

File: MOE Decon Technologies and Performance / Results of Model Decon Projects

[41] Life support team, Cabinet Office nuclear victims, Fukushima except decontamination project team, Japan Atomic Energy Research Development Organization, *Kawamata-Machi Sakashita District Except Decontamination Model Demonstration Project Progress Report,* 11

Appendix 1, except dyeing model projects for municipalities of results (details), 3/2013

[42] Ministry of the Environment, *Outlines of the Act on Special Measures concerning the Handling of Radioactive Pollution,* mm/dd/yyyy

[43] Ministry of the Environment, *Outlines of the Act on Special Measures concerning the Handling of Radioactive Pollution,* 3/5/2013

[44] Ministry of the Environment, *Interim Storage Facility*, 10/20/2011

[45] Ministry of the Environment, *Decontamination Plan Overview,* mm/dd/yyyy

[46] Ministry of the Environment, Japan, *Management of Off-site Waste Contaminated with Radioactive Materials Due to the Accident at Fukushima Nuclear Power Stations,* 11/28/2012

[47] Ministry of the Environment, Japan, *Management of off-site Waste Contaminated with Radioactive Materials Due to the Accident at Fukushima Nuclear Power Stations,* 11/28/2012

[48] Ministry of the Environment, Basic Policy on Interim Storage and Other Facilities Required for the Handling of the Environmental Pollution from Radioactive Materials Associated with the Accident at Tokyo Electric *Power Co.'s Fukushima Daiichi Nuclear Power Stations,* 10/29/2011

[49] Takuya NOMOTO, Koji ONO, Ministry of the Environment, Japan, *Management of Off-site Waste* Contaminated with Radioactive Materials due to the Accident at Fukushima Nuclear Power Stations, 11/28/2012

[50] Ministry of the Environment, [illegible]
(別紙２)

[51] Ministry of the Environment, *Efforts to Secure Interim Storage Facility,* mm/dd/yyyy

[52] Ministry of the Environment, *Schematic Diagram of a Temporary Storage Facility (Example)*, mm/dd/yyyy

[53] Hideki Kawamura, Obayashi Corporation, *Obayashi Activities on Environmental Remediation and Interim Storage Project,* 3/6/2013

[54] Radiation Monitoring Coordination Meeting, *Comprehensive Radiation Monitoring Plan (Provisional Translation),* 4/1/2012

[55] Radiation Monitoring Coordination Meeting, *Monitoring Surveys Incorporated into the Comprehensive Monitoring Plan (revised on April 1, 2012, at the Monitoring Coordination Meeting),* 4/1/2012

[56] Tadashi Inoue, Clean-up Committee of the Atomic Energy Society of Japan, *Overview of Environmental Contamination by Radioactivity*

Discharged from Fukushima Daiichi NPP, Japan/U.S. DOE Workshop on Remediation, Hanford, Washington, 2/13-15/2012

[57] Team in Charge of Assisting the Lives of Disaster Victims, Cabinet Office, Emergency Operation Center, MEXT, Ministry of the Environment, *Radiation Monitoring Action Plan for Supporting Restoration of the Former Emergency Evacuation Preparation Areas (Minami Soma city, Tamura city, Kawauchi village, Hirono town, and Naraha town),* 10/3/2011
File: Monitoring / houshasei_110719_m1

[58] *Drinking Water Radiological Substance Examination about Fukushima Prefecture*, 6/30/2011

[59] Ministry of Education, Culture, Sports, Science and Technology, *Renewal of the Website for Releasing Measurement Results of Air Dose Rates Nationwide and in Fukushima Prefecture (Real-time Distribution of Measurement Results at Monitoring Posts and under Real-time Radiation Dose Measurement System),* 5/14/2012

[60] Ministry of Education, Culture, Sports, Science and Technology, *Measures to be Taken by MEXT for Specific Spots Outside Fukushima Prefecture that Show Higher Radiation Doses than the Surrounding Areas,* 10/21/2011

[61] Team in Charge of Assisting the Lives of Victims around the Nuclear Power Plant Cabinet Office, Reconstruction Agency, Emergency Operation Center, MEXT, Ministry of the Environment, Fukushima Prefecture, *Radiation Monitoring Action Plan for Supporting the Return to/Restoration of Areas to which Evacuation Orders Are Ready to be Lifted (Minami Soma City, Tamura City, Kawauchi Village),* 4/18/2012
File: Monitoring / Monitoring Surveys 1

[62] *Monitoring Surveys Incorporated into the Co:,* mm/dd/yyyy
File: Monitoring / Monitoring Surveys 2

[63] *Monitoring Plan (Revised on April I, 2012. at the Monitoring Coordination Meeting),* 4/1/2012

[64] Task Force for the Reform of Nuclear Safety Regulations and Organizations, Cabinet Secretariat, Government of JAPAN, *Reform of Nuclear Regulation Organisation and System in Japan,* mm/dd/yyyy

[65] MEXT: Ministry of Education, Culture, Sports, Science and Technology, MLIT: Ministry of 1 and, Infrastructure, Transport and Tourism, MAFF: Ministry of Agriculture, forestry and Fisheries, ENV: Ministry of the Environment, JCG: Japan Coast Guard, JMA: Japan Meteorological Agency, 1 NERH: local Nuclear Emergency Response

Headquarters, TEPCO: Tokyo Electric Power Company, FEPC: The Federation of Electric Power Companies of Japan, Current Status of Environmental Radiation Monitoring Based on the *Comprehensive Radiation Monitoring Plan,* 2/28/13

[66] Radiation Monitoring Coordination Meeting 2/28/13 (NRA), Comprehensive Radiation Monitoring Plan *(Provisional translation),* 4/1/2012

[67] MEXT: Ministry of Education, Culture, Sports, Science and Technology, MLIT: Ministry of Land, Infrastructure, Transport and Tourism, MAf f: Ministry of Agriculture, forestry and fisheries, ENV: Ministry of the Environment, JCG: Japan Coast Guard, JMA: Japan Meteorological Agency, LNERH: Local Nuclear Emergency Response Headquarters, TEPCO: Tokyo Electric Power Company, FEPC: The federation of Electric Power Companies of Japan, *Current Status of Environmental Radiation Monitoring Based on the Comprehensive Radiation Monitoring Plan,* 4/1/2012

[68] MEXT: Ministry of Education, Culture, Sports, Science and Technology, MliT: Ministry of Land, Infrastructure, Transport and Tourism, MAFF: Ministry of Agriculture, Forestry and Fisheries, ENV: Ministry of the Environment, JCG: Japan Coast Guard, JMA: Japan Meteorological Agency, LNERH: Local Nuclear Emergency Response Headquarters, TEPCO: Tokyo Electric Power Company, FEPC: The Federation of Electric Power Companies of Japan, *Current Status of Environmental Radiation Monitoring Based on the Comprehensive Radiation Monitoring Plan,* 4/1/2012

[69] Team in charge of Assisting the Lives of Nuclear Disaster Victims, Ministry of Education, Culture, Sports, Science and Technology, *Plan to Conduct Detailed Monitoring in Restricted Area and Planned Evacuation Zone,* 6/13/2011

[70] Ministry of Education, Culture, Sports, Science and Technology, Fisheries Agency, Ministry of Land, Infrastructure, Transport and Tourism, Japan Coast Guard, Japan Meteorological Agency, Ministry of the Environment, Fukushima prefecture, Tokyo Electric Power Co. Inc., *Sea Area Monitoring Plan in FY20 12,* 3/30/2012

[71] Ministry of Education, Culture, Sports, Science and Technology, Fisheries Agency, Ministry of Land, Infrastructure, Transport and Tourism, Japan Coast Guard, Japan Meteorological Agency, Ministry of the Environment, Fukushima prefecture, Tokyo Electric Power Co. Inc., *Sea Area Monitoring Plan in FY2012,* 3/30/2012

[72] National Institute for Environmental Studies, 16-2 Onogawa, Tsukuba, Ibaraki 305-8506, Japan, *Fact Sheet,* mm/dd/yyyy

[73] N Masahiro Natsuhori, Japan Animal Referral Medical Center (JARMeC), *Regarding Countermeasures on the Impact of Contamination by Radioactive Materials on Animals Following the Nuclear Power Plant Disaster Caused by the Accident of Fukushima Daiichi Nuclear Power Plant and Countermeasures Against Contamination (Emergency Provisional Actions),* 4/1/2011

[74] Toshimasa Ohara, National Institute for Environmental Studies, *Introduction to National Institute for Environmental Study's Research Project on "Behaviour of Radionuclides in the Environment and Organisms,"* mm/dd/yyyy

[75] Ministry of the Environment, *Proposal Item from MOE,* mm/dd/yyyy

File: rad worker guidelines / 120625-01

[76] Guidelines for *Radiation Failure Prevention of Workers Engaged in the Decontamination Operations Overview,* 12/22/2011 With a base of 1222 item (vi). 6/15/2012 With base from 0615 amendment by no. 6.

File: rad worker guidelines / 120625-02

[77] *Guidelines for Radiation Preventing Worker Engaged in Specific Doses Below Business Overview,* (6/15/2012 With base from 0615, no. 6)

File: Remediation Strategy Progress and Status / Decontamination roadmap

[78] Ministry of the Environment, Japan, "*Decontamination Roadmap" for the Special Decontamination Areas*, 1/26/2012

File: Remediation Strategy Progress and Status / decontamination status

市町村除染地域における除染実施状況（平成25年1月末時点）

[79] [illegible]

[80] Ministry of the Environment, Japan, *The Long-term Strategy for Remediation,* 2/5/2013

[81] Ministry of the Environment, Fukushima Office, 特別地域内除染実施計画における実施対象区域について

[82] Ministry of the Environment, Japan, *Progress on Offsite Cleanup .Efforts in Japan,* 9/2012

File: Remediation Strategy Progress and Status/Remediation Strategy

[83] *Progress of Work in the Special Decontamination Area,* mm/dd/yyyy

[84] Decontamination Information Plaza, Fukushima Office for Environmental Restoration, Ministry of the Environment, *Status of the*

Contamination Caused by Fukushima Daiichi N PS Accident, its Health Effects and Decontamination Efforts, 2/2013

[85] Ministry of the Environment, *Policies for the Decontamination of Specific Areas Points for "Decontamination Roadmap,"* mm/dd/yyyy

[86] International Cooperation Office for Decontamination of Radioactive Materials, Ministry of the Environment, Japan, *Progress of Off-site Decontamination in Japan (Brief Overview),* 12/15/2012

[87] International Atomic Energy Agency, *Summary Report of the Preliminary Findings of the NE/NEFW /2011 ORIGINAL: English IAEA Mission on Remediation of Large Contaminated Areas Off-site the Fukushima Dai-ichi NPP, 7-15 October 2011, Japan,* 10/14/2011

[88] Mitsugu Kato, Japan Atomic Energy Agency, *(2) Report of the Results of the Decontamination Model Project, Overview of the Results of the Decontamination Model Project – Overview of the Results of Group B,* 3/26/2012

[89] Kazuki lijima, Japan Atomic Energy Agency, *(2) Report of the Results of the Decontamination Model Project, Overview of the Results of the Decontamination Model Project – Overview of the Results of Group C,* 3/26/2012

[90] Kaname Miyahara and Hiromitsu Saegusa, Japan Atomic Energy Agency, *Overview of the Results of the Decontamination Model Projects,* mm/dd/yyyy

[91] Yoichi Ito, Japan Atomic Energy Agency, *(2) Report of the Results of the Decontamination Model Projects - Overview of the Results of the Decontamination Model Projects – Role and Overview of the Model Projects,* 3/26/2012

[92] Topics Fukushima, JAEA, *Project Entrusted by the Ministry of the Environment –Creating Reports of Testing of 22 Decontamination Technologies,* 11/30/2012

[93] Ministry of the Environment, Japan, *Progress on Offsite Cleanup Efforts in Japan*, mm/dd/yyyy

[94] Japan Atomic Energy Agency, *Report of the Results of the Decontamination Model Projects,* 12/20/2012

[95] Ministry of the Environment, *Decontamination Model Projects in High Dose Area (Discussion Paper)*, mm/dd/yyyy

[96] Fukushima Office for Environmental Restoration, Tohoku Regional Environmental Office, Ministry of the Environment, *Advanced Works for Decontamination and Related Operations in the Town of Okuma for Fiscal Year 2012,* 10/11/2012

[97] Takuya NOMOTO, Koji ONO, Ministry of the Environment, Japan, *Management of Off-site Waste Contaminated with Radioactive Materials due to the Accident at Fukushima Nuclear Power Stations*, 11/28/2012

[98] Ministry of the Environment, *Answers to Your Frequently Asked Questions on the Antiterrorism Law Dealing with Radioactive Substance Pollution Waste Related,* 6th Edition, 7/2012

[99] Ministry of the Environment, 廃棄物関係ガイドライン
事故由来放射性物質により汚染された
廃棄物の処理等に関するガイドライン
平成23年12月　第1版
File: waste guidelines / haikibutsu-gl01_verh2412

[100] Ministry of the Environment, 第一部,
汚染状況調査方法
ガイドライン
12/2012 2
File: waste guidelines / haikibutsu-gl02_verh2412

[101] Ministry of the Environment, 第二部
特定一般廃棄物・
特定産業廃棄物関係
ガイドライン
12/2012
File: waste guidelines / haikibutsu-gl03_verh2412

[102] Ministry of the Environment, 第二部
指定廃棄物関係
ガイドライン
12/2012
File: waste guidelines / haikibutsu-gl04_verh2412

[103] Ministry of the Environment, 第二部
除染廃棄物関係
ガイドライン
12/2012
File: waste guidelines / haikibutsu-gl05_ver1

[104] Ministry of the Environment, 第二部

放射能濃度等測定方法
ガイドライン
平成23年12月 第1版

File: waste guidelines / haikibutsu-gl06_ver1

[105] Ministry of the Environment,
特定廃棄物関係
ガイドライン

[106] Kenkichi Ishigure, Decontamination Technology Demonstration Promotion Committee, *Overview of Decontamination Projects,* mm/dd/yyyy

[107] Takeshi SEKIYA, International Coordination Unit, Taskforce for Radioactive Environmental Pollution, Ministry of the Environment, Japan, *Overview of Japan's Offsite Cleanup Efforts,* mm/dd/yyyy

[108] Ministry of the Environment, Japan, *Health Management of Radiation Exposure in Fukushima,* mm/dd/yyyy

[109] *Act on Special Measures concerning the Handling of Environment Pollution by Radioactive Materials Discharged by the NPS Accident Associated with the Tohoku District- Off the Pacific Ocean Earthquake That Occurred on March 11, 2011,* 11/11/2011

[110] *Questions & Answers,* 3/6/2013

[111] Steve Rima, AMEC Environment & Infrastructure, *Cleanup Associated with Fukushima Incident – Radiological Survey & Soil Sorting for Waste Minimization,* mm/dd/yyyy

[112] Ministry of the Environment, "Decontamination Roadmap" for the Special Decontamination Areas, 1/26/2012

[113] International Atomic Energy Agency, Summary Report of the Preliminary Findings of the IAEA Mission on remediation of large contaminated areas off-site the Fukushima Dai-ichi NPP, 7–15 October 2011, Japan, 10/14/2011

File: MAFF

[114] Ministry of the Environment, *Decontamination Model Projects in High Dose Area (Discussion Paper)*, mm/dd/yyyy

[115] Ministry of the Environment, *List of Questions from the MOE*, mm/dd/yyyy

APPENDIX 3. BIBLIOGRAPHY OF DOCUMENTATION SUBMITTED TO GOJ MOE

Radiation Protection			
Material		Description	Link
1.	International Commission on Radiological Protection (ICRP). **Application of the Commission's Recommendations to the Protection of People Living in Long-term Contaminated Areas After a Nuclear Accident or a Radiation Emergency**	In this report, the Commission provides guidance for the protection of people living in long-term contaminated areas resulting from either a nuclear accident or a radiation emergency. The report considers the effects of such events on the affected population.	http://www.icrp.org/docs/P111(Special%20Free%20Release).pdf
2.	United States Department of Energy (DOE). **The CERCLA process**	CERCLA advisory board meeting example presentation	electronic file available upon request
3.	United States Department of Energy (US DOE). **Preliminary Report on Operational Guidelines Developed for Use in Emergency Preparedness and Response to a Radiological Dispersal Device Incident**	This report consolidates preliminary operational guidelines and descriptions of their derivation, all ancillary work products, and a companion software tool (RESRAD-RDD) that facilitates their implementation into one reference source document. It is intended for use and comment by Federal agencies, State and local governments, and emergency planning and response subject matter experts regarding emergency preparedness planning and response initiatives for an RDD incident.	http://www.epa.gov/radiation/docs/er/ogt_manual_doe_hs_0001_2_24_2009.pdf
4.	United States Department of Energy (US DOE). **RESRAD Family Codes**	In December 2007, new versions of RESRAD (V.6.4) and RESRAD-BUILD (V.3.4) codes were released. RESRAD is a computer model code designed to estimate radiation doses and risks from RESidual RADioactive materials, sponsored by the Office of Health, Safety and Security and the Office of Environmental Management, with support from the U.S. Nuclear Regulatory Commission. This family of codes was developed by Argonne National Laboratory (ANL); code and version control are currently maintained by the Department of Energy (DOE) through ANL.	http://web.ead.anl.gov/resrad/home2/
5.	United States Department of Energy Office of Environmental Management.	The Savannah River Site (SRS) -Citizens Advisory Board (CAB) provides the Assistant Secretary for Environmental Management and designees with advice, information, and recommendations on issues	http://www.srs.gov/general/outreach/srs-cab/srs-cab.html

Radiation Protection

	Material	Description	Link
	Savannah River Site -Citizens Advisory Board.	affecting the EM program. Among those issues are clean-up standards and environmental restoration; waste management and disposition; stabilization and disposition of non-stockpile nuclear materials; excess facilities; future land use and long-term stewardship; risk assessment and management; and clean-up science and technology activities. The board's membership is carefully considered to reflect a full diversity of viewpoints in the community and region. Board members are composed of people who are directly affected by DOE site clean-up activities.	
6.	United States Department of Homeland Security/Federal Emergency Management Agency (US DHS/ FEMA. **Planning Guidance for Protection and Recovery Following Radiological Dispersal Device (RDD) and Improvised Nuclear Device (IND) Incidents. 2008.**	The Department of Homeland Security (DHS) is issuing final guidance entitled, ''Planning Guidance for Protection and Recovery Following Radiological Dispersal Device (RDD) and Improvised Nuclear Device (IND) Incidents'' (the Guidance).	http://www.epa.gov/radiation/docs/er/fema-FR-PlanGuidProteRec-RDD-IND-Incident.pdf
7.	United States Environmental Protection Agency (EPA). Dose Compliance Concentration (DCC) calculators	To address environmental standards that are expressed in terms of millirem per year, an approach similar to that taken for calculation of PRGs was also used to calculate soil "compliance concentrations" based upon various methods of dose calculation in another EPA tool, the "Dose Compliance Concentrations", or DCC calculator. The DCC calculator equations are identical to those in the PRG for Radionuclides, except that the target dose rate (ARAR based) is substituted for the target cancer risk (1×10^{-6}), the period of exposure is one year to indicate year of peak dose, and a dose conversion factor (DCF) will be used in place of the slope factor. EPA developed two other electronic calculators. These are the Radionuclide Building Dose Cleanup Concentrations (BDCC) and the Radionuclide Outside Hard Surfaces Dose Cleanup Concentrations (SDCC) electronic calculators. Both of these ARAR dose calculators are set up in a similar manner to the BPRG and SPRG calculators.	http://www.epa.gov/superfund/health/contaminants/radiation/radrisk .htm

(Continued)

Radiation Protection			
	Material	**Description**	**Link**
8.	United States Environmental Protection Agency (EPA). Preliminary Remediation Goals (PRG) calculators	EPA has developed several tools to calculate Preliminary Remediation Goals (PRGs) cancer risk-based concentrations, derived from standardized equations combining exposure information assumptions with EPA toxicity data. EPA has developed a PRG for Radionuclides electronic calculator, known as the Rad PRG calculator. This electronic calculator presents risk- based standardized exposure parameters and equations that should be used for calculating radionuclide PRGs for residential, commercial/industrial, and agricultural land use exposures, tap water and fish ingestion exposures. The calculator also presents PRGs to protect groundwater. The Preliminary Remediation Goals for Radionuclides in Buildings (BPRG) electronic calculator was developed to help standardize the evaluation and cleanup of radiologically contaminated buildings at which risk is being assessed for occupancy. BPRGs are radionuclide concentrations in dust, air and building materials that correspond to a specified level of human cancer risk. The Preliminary Remediation Goals for Radionuclides in Outside Surface SPRG calculator addresses hard outside surfaces such as building slabs, outside building walls, sidewalks and roads. SPRGs are radionuclide concentrations in dust and hard outside surface materials. The BPRG and SPRG calculators include both residential and industrial/commercial exposure scenarios.	http://www.epa.gov/superfund/health/contaminants/radiation/ra drisk.htm
9.	United States Environmental Protection Agency (EPA)**. A guide to selecting superfund remedy actions**	CERCLA remedy selection guidance document	http://www.epa.gov/superfund/community/relocation/gui_sel1.pdf
10.	United States Environmental Protection Agency (EPA). CERCLA superfund orientation manual	The "CERCLA/Superfund Orientation Manual" serves as a program orientation guide and reference document, and it is designed to assist EPA and State personnel involved with hazardous waste remediation, emergency response, and chemical and emergency preparedness. The	http://www.epa.gov/superfund/policy/remedy /pdfs/542r-92005-s.pdf

Radiation Protection

	Material	Description	Link
		Manual describes the organizational and operational components of the EPA Superfund Program.	
11.	United States Environmental Protection Agency (EPA). **Draft PAG Manual**. (Proposed revision/update to United States Environmental Protection Agency (EPA). Manual of Protective Action Guides and Protective Actions for Nuclear Incidents, #13 below)	Protective Action Guides (PAGs) suggest precautions that state and local authorities can take during an emergency to keep people from receiving an amount of radiation that might be dangerous to their health. EPA developed the PAG Manual to provide guidance on actions to protect the public, such as having people evacuate an area or stay indoors.	http://www.epa.gov/radiation/docs/er/pag-manual-interim-public-comment-4-2-2013.pdf
12.	United States Environmental Protection Agency (EPA). **Key Principles of Remedy Selection**	List of documents for key principles of remedy selection for EPA Superfund sites	http://www.epa.gov/superfund/policy/remedy/sfremedy/remedies/principles.htm
13.	United States Environmental Protection Agency (EPA). **Manual of Protective Action Guides and Protective Actions for Nuclear Incidents**	Protective Action Guides (PAGs) help state and local authorities make radiation protection decisions during emergencies. EPA developed the PAG Manual to provide guidance on actions to protect the public.	http://www.epa.gov/radiation/docs/er/400-r-92-001.pdf
14.	United States Environmental Protection Agency (EPA). **Radiation Protection Guidance to Federal Agencies for Occupational Exposure; Approval of Environmental Protection Agency Recommendations**	This memorandum transmits recommendations that would update previous guidance to Federal agencies for the protection of workers exposed to ionizing radiation. These recommendations were developed cooperatively by the Nuclear Regulatory Commission, the Occupational Safety and Health Administration, the Mine Safety and Health Administration, the Department of Defense, the Department of Energy, the National Aeronautics and Space Administration, the Department of Commerce, the department of Transportation, the Department of Health and Human Services, and the environmental Protection Agency. In addition, the NCRP, the National Academy of Sciences (NAS), the conference of Radiation Control Program Directors (CRCPD) of the stats, and the Health Physics Society were consulted during the development of this guidance.	http://www.epa.gov/radiation/docs/federal/5 2-fr-2822.pdf
15.	United States Environmental Protection Agency (EPA). **Radiation Task Force Leader Training**	it is an example of how to use ALL non-rad science people in assisting with a rad incident. In only 2 weeks, non-rad people, with science backgrounds, can be brought up to a level where they can conduct	http://www.trainex.org/offeringslist.cfm?courseid=880

(Continued)

Radiation Protection			
Material		Description	Link
		radiation surveys, release previously contaminated things for use (under supervision of an Health Physicist), take air samples, count contamination swipes and air samples and assist the "ground zero" workers in exiting the hotter areas (decon line ops), under supervision of an Health Physicist.	
16.	United States Environmental Protection Agency (EPA). **RCRA online training**	This website consolidates Resource Conservation and Recovery Act (RCRA) -related online courses, seminars, webinars, podcasts, and videos that are posted throughout the Wastes website. Both introductory and more advanced courses are included for federal and state regulators, the regulated community, organizations, associations, and even consumers who are interested in environmental laws and regulations and their implementation.	http://www.epa.gov/epawaste/education/train .htm
17.	United States Environmental Protection Agency (EPA). **Rules of Thumb for Superfund Remedy Selection**	CERCLA remedy selection guidance document	http://www.epa.gov/superfund/policy/remedy /rules/rulesthm.pdf
18.	United States Environmental Protection Agency (EPA)/ Interstate Technology Regulatory Council (ITRC). **Radiation Risk/Dose Assessment: Updates and Tools**	The ITRC Radionuclides Team's Determining Cleanup Goals at Radioactively Contaminated Sites: Case Studies (RAD-2, 2002) examines the factors influencing variations in cleanup level development at various radioactively contaminated sites and underscores the need for training to enhance consistency in radiation risk assessment application. The document also acknowledges the differences between the 'dose approach' used at some sites and EPA's 'risk-based approach.' Since most radioactively contaminated DOE and DOD sites are developing cleanup goals under CERCLA authority, there is a need for training that clarifies the variations between these approaches and elaborates on the methodology used to develop risk-based remediation goals. This training course has been collaboratively developed by the ITRC Radionuclides Team and EPA's Superfund Office to meet these needs. The focus of this	http://www.clu-in.org/conf/itrc/rads_051507/

Radiation Protection			
Material		Description	Link
		training is EPA's new radiation risk assessment tools, which can facilitate better decision making for accelerated cleanups.	
19.	United States Environmental Protection Agency (EPA)/ Interstate Technology Regulatory Council (ITRC). **Radiation Site Cleanup: CERCLA Requirements and Guidance**	The ITRC Radionuclides Team's "Determining Cleanup Goals at Radioactively Contaminated Sites: Case Studies" (RAD-2, April 2002) examines the factors influencing variations in cleanup level development at various radioactively contaminated sites and underscores the need for training to enhance consistency in remedy selection for radiological contaminants. Since most radioactively contaminated DOE and DOD sites are developing cleanup goals under CERCLA authority, there is a need for training that elaborates on the methodology used to select remedies under EPA's approach for CERCLA sites.	http://www.clu-in.org/conf/itrc/radscleanup_060507/
20.	United States Federal Radiological Monitoring and Assessment Center (FRMAC). **Assessment Manual Volume 1 Overview and Methods**	This FRMAC Assessment Manual has been prepared by representatives of those Federal and State agencies that can be expected to play the major roles during a radiological emergency. Federal Agencies include: the National Nuclear Security Administration (NNSA), the Nuclear Regulatory Commission (NRC), the Environmental Protection Agency (EPA), the Department of Agriculture (USDA), the Food and Drug Administration (FDA), and the Centers for Disease Control (CDC). This final manual was reviewed by experts from across the community and their input has been incorporated.	http://www.nv.doe.gov/library/publications/frmac/FRMAC%20Division/FRMAC%20Assessment/FRMAC%20Assessment%20Manual%20Vol%201/FRMAC%20Assessment%20Manual,%20Volume%201%20-%20Overview%20and%20Methods.pdf
21.	United States National Council on Radiation Protection and Measurements (NCRP). Decision Making for Late-Phase Recovery **from Nuclear or Radiological Incidents (DRAFT)**	Preparing against industrial nuclear or radiological incidents has been going on in the United 305 States for several decades. Most of the information has drawn upon the experiences gained from 306 the past incidents that have occurred around the world. And specific improvements have often 307 been made following a major incident of concern. To facilitate protection of the public from 308 potential radiological exposures during incidents, the Protective Action Guide (PAG) was 309 developed by the U.S. Environmental Protection Agency (EPA) in the 1970s.	http://www.ncrponline.org/Docs_in_Review/NCRPM1302.pdf

(Continued)

Radiation Protection			
Material		Description	Link
		Following the TMI 310 nuclear accident, a streamlined federal leadership hierarchy assigned EPA the task of 311 establishing its guidance, Manual of Protective Action Guides and Protective Actions for 312 Nuclear Incidents (EPA, 1992), for radiological response planning in 1992 by taking into 313 account the lessons learned.	
Decontamination Methods/Technologies			
1.	Denham, Miles. ***Returning ^{137}Cs and ^{90}Sr Contaminated Land to Productive Use.*** Savannah River National Laboratory. 2013.	Proposal to DOE-EM International Program for joint project with International Radioecology, Slavutych, Ukraine, and Ministry of Emergency Situations of the Republic of Belarus—Pollessie State Radiation-Ecological Reserve.	Electronic file provided
2.	Farfan, Eduardo, and Rusty Coleman. "**Comparison of Performance and Cost of GrayQb to other 3D Radiation Mapping Devices**." Savannah River National Laboratory. 2013.		Electronic file provided
3.	Farfan, Eduardo, and Rusty Coleman. **Note on Toshiba and Hitachi Gamma Camera Systems. Savannah** River National Laboratory. 2013.		Electronic file provided
4.	Interstate Technology and Regulatory Council "Decontamination and Decommissioning of Radiologically-Contaminated Facilities"	This Internet-based training was developed collaboratively by EPA with the Radionuclides Team of the Interstate Technology and Regulatory Council (ITRC), a state-led coalition working together with industry and stakeholders to achieve regulatory acceptance of environmental technologies.	http://www.clu-in.org/conf/itrc/radsdd_040308/
5.	Interstate Technology and Regulatory Council Decontamination and Decommissioning of Radiologically Contaminated Facilities	This document compiles knowledge and experience acquired in recent years from facilities that have completed a D&D process, providing guidance on D&D to regulators, the public, project managers, cleanup contractors, and technology providers.	http://www.itrcweb.org/Guidance/ListDocuments?TopicID=21&SubTopicID=24

Decontamination Methods/Technologies			
Material		Description	Link
6.	United States Environmental Protection Agency Technology Reference Guide for Radiologically Contaminated Surfaces	This guide identifies various surface decontamination technologies that can be used to remove radiation contaminates from building, structure, and equipment surfaces. Technology profiles provide information on both chemical and physical contamination technologies, including target contaminants, waste management issues, operating characteristics, and associated cost. The information presented in this guide allows technologies to be compared for site- specific use.	http://www.epa.gov/radiation/docs/cleanup/4 02-r-06-003.pdf
7.	United States Environmental Protection Agency Technology Reference Guide for Radioactively Contaminated Media	This guide references various technologies that can be used to treat radioactively contamination present in liquid media, including ground water, surface water, and waste water, and solid media, including soil, sediment, and solid waste. Information concerning each of 21 applied technologies is presented in technology profiles, which can be used to compare technologies for site-specific application. Five emerging technologies are also profiled in this guide.	http://www.epa.gov/superfund/health/contaminants/radiation/pdfs/technology_ref_guide_for_contaminated_media.pdf
8.	Westinghouse Savannah River Company. ***Soil and Groundwater Closure Projects: Technology Descriptions.*** WSRC-RP-99-4015. Revision 7.1. January 2007	This book summarizes those technologies that have been implemented to facilitate process improvements, operable unit characterizations, and remedial actions at the Savannah River Site (SRS) over the last 15 years. Overall, 105 new technologies have been applied to the environmental program at the Savannah River Site (SRS). Many of these technologies have been redeployed for use at other operable units; and in a number of cases, some technologies have been institutionalized as the standard mode of operations. The following figure depicts the number of new technology and technology redeployments from 1996 to 2006. The number of technologies redeployed is accounted by a reuse following the initial deployment. beginning in the mid-1990s, many of the technologies deployed were new as the program was in its infancy, moving from document production to field work upon regulatory approval. With more field activities ongoing, lessons learned and implementation of numerous technologies, allowed for many technologies that were initially deployed to be redeployed. This is evident as the environmental program matured into the early 21st	http://www.srs.gov/general/programs/soil/gen/sgcp_tech_descriptions.pdf

(Continued)

Decontamination Methods/Technologies			
Material		Description	Link
		century. After FY03, the number of field activities diminished resulting in a reduction in the number of technologies deployed. However, the program is still active in all aspects of environmental restoration and is actively promoting the use of new technologies and redeployment of technologies that expedite field work and operable unit closure while reducing cost and improving schedule efficiencies.	
Waste Management System			
1.	Bellamy, Steve. ***Radioactive Material Transportation Packaging Technology & Pressurized Systems***.	Overview of the engineering group at SRNL with mission to design radioactive materials (RAM) packaging, perform RAM package analysis and regulatory support to DOE complex.	Presentation provided to MOE on electronic media
2.	**Gelder, Lawrence F. *Considerations and Technologies for Nuclear Materials Packaging.***	SRNL designs and develops packages for radioactive materials for the U.S. Departments Energy, Homeland Security, and the Federal Bureau of Investigation. These packages are tested and/or analyzed, and meet the certification requirements of the U.S. Nuclear Regulatory Commission.	Presentation provided to MOE on electronic media
3.	International Atomic Energy Agency (IAEA). Technical Report 330. **Disposal of Waste from the Cleanup of Large Areas Contaminated as a Result of a Nuclear Accident.**	The problem of ground contamination in the case of a severe nuclear accident is one of special concern. This report gives guidance on planning and management of safe transportation and disposal of large volumes of contaminated materials, with the objective of minimizing the consequences of such an accident. It is closely linked to Technical Reports Series No. 300, Cleanup of Large Areas Contaminated as a Result of a Nuclear Accident (1989), and No. 327, Planning for Cleanup of Large Areas Contaminated as a Result of a Nuclear Accident (1991).	http://www-pub.iaea.org/mtcd/publications/PubDetails.asp?pubId=1434
4.	United States Department of Energy (US DOE). **The CERCLA process.**	CERCLA advisory board meeting example presentation	electronic file
5.	United States Environmental Protection Agency (US EPA). Letter to Idaho State Senator Clint Stennett.	Letter from Robert Perciasepe, then-EPA Assistant Administrator of the Office of Air and Radiation, and Timothy Fields, Jr., then-Assistant Administrator of the Office of Solid Waste and Emergency	Electronic file

Waste Management System			
Material		Description	Link
		Response, to The Honorable Clint Stennett, Minority Leader, Idaho State Senate, June 26, 2000. Describes conditions for disposal of radioactive waste from a CERCLA site at a non-NRC licensed facility.	
6.	United States Environmental Protection Agency (US EPA). Statement from OSWER DAA to Senate Environment and Publics Work Committee.	Statement of Michael Shapiro, Principal Deputy Assistant Administrator, Office of Solid Waste and Emergency Response, before the U.S. Senate Committee on Environment and Public Works (July 25, 2000).	Electronic file
7.	United States Environmental Protection Agency (US EPA). Fact Sheet on Evapotranspiration Cover Systems for Waste Containment	The fact sheet provides a summary of ET technical issues, including design considerations, performance monitoring, cost, technology status, and potential limitations on use.	http://www.clu-in.org/download/remed/epa542f11001.pdf
8.	United States Environmental Protection Agency (US EPA). Evapotranspiration Covers	This website provides information and links about Evapotranspiration Covers.	http://www.clu-in.org/products/evap/add/default.cfm
9.	United States Environmental Protection Agency (US EPA). ***A Guide to Selecting Superfund Remedy Actions***.	CERCLA remedy selection guidance document	http://www.epa.gov/superfund/community/relocation/gui_sel1.pdf
10.	United States Environmental Protection Agency (US EPA). ***CERCLA Superfund Orientation Manual.***	The "CERCLA/Superfund Orientation Manual" serves as a program orientation guide and reference document, and it is designed to assist EPA and State personnel involved with hazardous waste remediation, emergency response, and chemical and emergency preparedness. The Manual describes the organizational and operational components of the EPA Superfund Program.	http://www.epa.gov/superfund/policy/remedy /pdfs/542r-92005-s.pdf
11.	United States Environmental Protection Agency (US EPA). **Key Principles of Remedy Selection**.	List of documents for key principles of remedy selection for EPA Superfund sites	http://www.epa.gov/superfund/policy/remedy/sfremedy/remedies/principles.htm
12.	United States Environmental Protection Agency (US EPA). **RCRA online training**	This website consolidates Resource Conservation and Recovery Act (RCRA) -related online courses, seminars, webinars, podcasts, and videos that are posted throughout the Wastes website. Both introductory and more advanced courses are included for federal and state regulators, the regulated community, organizations, associations,	http://www.epa.gov/epawaste/education/train .htm

(Continued)

Waste Management System			
Material		Description	Link
		and even consumers who are interested in environmental laws and regulations and their implementation.	
13.	United States Environmental Protection Agency (US EPA). **Rules of Thumb for Superfund Remedy Selection.**	CERCLA remedy selection guidance document	http://www.epa.gov/superfund/policy/remedy/rules/rulesthm.pdf
14.	United States Environmental Protection Agency (US EPA)/ Interstate Technology Regulatory Council Radiation ***Site Cleanup: CERCLA Requirements and Guidance.***	The ITRC Radionuclides Team's "Determining Cleanup Goals at Radioactively Contaminated Sites: Case Studies" (RAD-2, April 2002) examines the factors influencing variations in cleanup level development at various radioactively contaminated sites and underscores the need for training to enhance consistency in remedy selection for radiological contaminants. Since most radioactively contaminated DOE and DOD sites are developing cleanup goals under CERCLA authority, there is a need for training that elaborates on the methodology used to select remedies under EPA's approach for CERCLA sites.	http://www.clu-in.org/conf/itrc/radscleanup_060507/
15.	United States Environmental Protection Agency (US EPA)/ Interstate Technology Regulatory Council. ***Radiation Risk/Dose Assessment: Updates and Tools.***	The ITRC Radionuclides Team's Determining Cleanup Goals at Radioactively Contaminated Sites: Case Studies (RAD-2, 2002) examines the factors influencing variations in cleanup level development at various radioactively contaminated sites and underscores the need for training to enhance consistency in radiation risk assessment application. The document also acknowledges the differences between the 'dose approach' used at some sites and EPA's 'risk-based approach.' Since most radioactively contaminated DOE and DOD sites are developing cleanup goals under CERCLA authority, there is a need for training that clarifies the variations between these approaches and elaborates on the methodology used to develop risk-based remediation goals. This training course has been collaboratively developed by the ITRC Radionuclides Team and EPA's Superfund Office to meet these needs. The focus of this training is EPA's new radiation risk assessment tools, which can facilitate better decision making for accelerated cleanups.	http://www.clu-in.org/conf/itrc/rads_051507/

Environmental monitoring			
	Material	Description	Link
1.	Abraham, J. P., F. W. Whicker, T. G. Hinton, D. J. Rowan. **"Inventory and spatial pattern of ^{137}Cs in a pond: a comparison of two survey methods."** *Journal of Environmental Radioactivity*. Vol. 51, No. 2, 2000, pp. 157-171.	The sediments of Pond A, a former Savannah River Site cooling pond for R-reactor, were contaminated with ^{137}Cs between 1954 and 1964. Pond A is unique because it is very shallow and contains an extremely high density of aquatic vegetation and old, undecomposed tree stumps which modify normal sedimentation processes and cause special radiological characterization challenges. To determine the most efficient technique for estimating inventory and spatial patterns of ^{137}Cs, we measured exposure rates at 124 sediment surface locations with two types of thermoluminescent dosimeters (TLDs) and compared them to estimates calculated from ^{137}Cs measured in 58 extracted sediment cores. The mean net exposure rate ($1SEM) measured at the sediment water interface with a UD-802 multielement TLD (differentially shielded lithium borate and calcium sulfate) was 40$4 lR h~1, while the corresponding value measured with a CaF 2 TLD was 64$10 lR h~1. Both sets of TLD measurements were found to correlate well with each other (R2"0.88, p(0.001), and moderately well with theoretical calculations derived from ^{137}Cs activity concentrations measured in sediment cores (R2'0.50). The corresponding mean exposure rate calculated from the sediment data, 69$10 lR h~1, was likely an over-estimate resulting from the core sampling bias created by the large number of tree stumps. Overall, peak ^{137}Cs activity occurred at&2}4 cm depth in the sediment cores, with 99% in the top 20 cm of sediment. The total ^{137}Cs inventory of Pond A was estimated as 4.1$0.5]1010 Bq, with most activity located in the deeper portions. Approximately 1% of the ^{137}Cs activity thought to have been released by RReactor can be accounted for in Pond A, with an additional 53% estimated from other work to be in the much larger Pond B, and Par Pond, located further down the drainage. However, the mean deposition in Pond A (7.9]109 Bq ha~1) was higher than either Pond B (4.0]109 Bq ha~1), or Par Pond (1.4]109 Bq ha~1). It was concluded that, although the TLD method was more efficient and could employ more sampling locations to estimate spatial pattern, a reasonable amount of coring was essential to determine	http://ac.elscdn.com/S0265931X00000527/1-s2.0-S0265931X00000527-main.pdf?_tid=23f34d3a-8ffc-11e2-9304-00000aacb362&acdnat=1363632567_5e38b237740531df3a98c6a7568cc7b0

(Continued)

Environmental monitoring			
Material		Description	Link
		depth distribution, radionuclide composition, and to interpret the TLD data. Optimal estimation and characterization efficiency can benefit from simultaneous application of both techniques.	
2.	Bertsch, P. M. (Savannah River Ecology Laboratory). **"Cesium-^{137} in floodplain sediments of the Lower Three Runs Creek on the DOE Savannah River Site.**" *Journal of Radioanalytical and Nuclear Chemistry*. Volume 264, Issue 2, May 2005, pp. 481-488.	The legacy of nuclear weapons production has resulted in vast tracks of land contaminated with fission products, mainly <>^{137}</>Cs, and at the U.S. Department of Energy's Savannah River Site (SRS) alone there is over 120 km<>2</> of land contaminated with low-levels of<> ^{137}</>Cs. Soils on the SRS are highly weathered and dominated by sand-sized quartz grains with the clay fraction consisting primarily of kaolinite and crystalline and poorly crystalline iron oxides (<5%). Our results showed that the majority of <>^{137}</>Cs in the Lower Three Runs Creek floodplain were retained in the sand-sized fraction (>52 mm) of the soil. Frayed edge site measurements were performed in order to probe the interaction between <>^{137}</>Cs and the sand fraction, with the results indicating that the vast majority of the <>^{137}</>Cs was strongly retained and existed in the residual fraction. These results prompted examination into the mineralogy of the soils in a hope to elucidate the mechanisms of <>^{137}</>Cs retention by the sand fraction. The results from this study provide new evidence for selective retention of <>^{137}</>Cs in larger-grained particles than previously demonstrated.	http://link.springer.com /article/10.1007/s1096 7-005-0741-6
aia3.	Carlton, W. H. ; Bauer, L. R., Evans, A. G.; Geary, L. A.; Murphy, C. E. Jr.; Pinder, J. E., Strom, R. N. (Westinghouse Savannah River Company). **"Cesium in the Savannah River Site environment."** WSRC-RP-92-250. March 1, 1992.	Cesium in the Savannah River Site Environment is published as a part of the Radiological Assessment Program (RAP). It is the fourth in a series of eight documents on individual radioisotopes released to the environment as a result of Savannah River Site (SRS) operations. The earlier documents describe the environmental consequences of tritium, iodine, and uranium. Documents on plutonium, strontium, carbon, and technetium will be published in the future. These are dynamic documents and current plans call for revising and updating each one on a two-year schedule. Radiocesium exists in the environment as a	http://www.osti.gov/bridge/product.biblio.jsp?query_id=0 &page=0&o sti_id=6914212&Row=0 &formname=basicsearc h.jsp

Environmental monitoring			
Material		Description	Link
		result of above- ground nuclear weapons tests, the Chernobyl accident, the destruction of satellite Cosmos 954, small releases from reactors and reprocessing plants, and the operation of industrial, medical, and educational facilities. Radiocesium has been produced at SRS during the operation of five production reactors. Several hundred curies of [sup ^{137}]Cs was released into streams in the late 50s and 60s from leaking fuel elements. Smaller quantities were released from the fuel reprocessing operations. About 1400 Ci of [sup ^{137}]Cs was released to seepage basins where it was tightly bound by clay in the soil. A much smaller quantity, about four Ci. was released to the atmosphere. Radiocesium concentration and mechanisms for atmospheric, surface water, and groundwater have been extensively studied by Savannah River Technology Center (SRTC) and ecological mechanisms have been studied by Savannah River Ecology Laboratory (SREL). The overall radiological impact of SRS releases on the off-site maximum individual can be characterized by total doses of 033 mrem (atmospheric) and 60 mrem (liquid), compared with a dose of 12,960 mrem from non-SRS sources during the same period of time. Isotope [sup ^{137}]Cs releases have resulted in a negligible risk to the environment and the population it supports.	
4.	Interstate Technology and Regulatory Council. Real-Time Measurement of Radionuclides in Soil: Technology and Case Studies	A technology overview document to educate regulators, contractors, site owners, stakeholders, and others involved in site cleanup decisions about the benefits of a streamlined data collection approach that has proven effective at radionuclide contaminated sites and may prove effective at other types of sites as well. The document describes the available technologies including benefits and limitations, processes for utilizing the technologies, and relevant site-specific experiences in implementing these technologies.	http://www.itrcweb.org/Guidance/ListDocuments?TopicID=21&SubTopicID=24
5.	Interstate Technology and Regulatory Council. Real-Time Measurement of Radionuclides in Soil	This training provides information on the basics of real-time measurement systems (detector types and platforms, location control and mapping technologies, surface and subsurface applications and limitations), how the technologies and data are used (characterization, remediation and closure, decision support, sources and types of	http://www.clu-in.org/conf/itrc/radsrealtime_102808/

(Continued)

Environmental monitoring			
Material		**Description**	**Link**
		uncertainty), acceptance issues (QA/QC, decision framework, uncertainty), and case studies. The purpose is to provide a solid background understanding of the technology itself and the context within which it is used.	
6.	Mohler, H. J., F. W. Whicker, T. G. Hinton. "Temporal trends of ^{137}Cs in an abandoned reactor cooling reservoir." Journal of Environmental Radioactivity. Vol. 37, No. 3, 1997, pp. 251-268.	A comprehensive resampling study was initiated for the purpose cf gaining insight in to ^{137}Cs mobility and retention in Pond B, an abandoned reactor cooling reservoir at the United States Department of Energy's Savannah River Site in Aiken, South Carolina. Measurements made during this study were compared to those made ten years earlier. Cesium-^{137} inventories were estimated in water, sediment, and biotic components. The total measured ^{137}Cs inventory in Pond B decreased,from 46xlO"Bq in 1984 to 2.3xlO"Bq in 1994. This decline was largely driven by the decrease measured in the sediment inventory, which approximates the total inventory. The results suggest a IO-year effective half-time for 'j7Cs in Pond B, which is significantly more rapid than the 28-year estimate made by Whicker et al. (1990). However, it is likely that the water turnover rate between 1984 and 1994 may have been higher than the rate of 0.3 year-' used for this estimate. Concentration ratios for the sediment and biota were very similar for both studies. A general trend of i37Cspenetration into the sediment profile and of sediment transport to deeper water was observed.	http://ac.els-cdn.com/S0265931X97000167/1-s2.0-S0265931X97000167-main.pdf?_tid=2aa9c3d a-8ffb-11e2-9a42-00000aab0f6c&acdnat=1363632149_8f595977 206540e8288e8cbee4f 7745f
7.	Pacific Northwest National Laboratory (PNNL). **Hanford On-line Environmental Information Exchange (PHOENIX)**	PHOENIX provides web-based GIS access to Hanford Site monitoring data bases including Hanford Environmental Information System and Tank Waste Information Network System. PHOENIX provides access to multiple data bases and allows data visualization through GIS maps, data tables, and charts.	http://phoenix.pnnl.gov
8.	Pacific Northwest National Laboratory (PNNL). **Visual Sample Plan (VSP)**	VSP is a software tool that supports the development of a defensible sampling plan based on statistical sampling theory and the statistical analysis of sample results to support confident decision making. VSP	www.vsp.pnl.gov

Environmental monitoring			
Material		Description	Link
		couples site, building, and sample location visualization capabilities with optimal sampling design and statistical analysis strategies. VSP supports the DQO process.	
9.	United States Department of Energy (US DOE). **Scientific Opportunities for Monitoring at Environmental Remediation Sites (SOMERS)**	DOE's vision for advancing monitoring through an integrated systems-based approach. This document identifies in detail scientific and technical challenges and opportunities associated with systems-based monitoring at DOE sites.	http://www.pnnl.gov/main/publications/external/technical_reports/PNNL-2[137]9.pdf
10.	United States Environmental Protection Agency (US EPA). **In-Situ Gamma-Ray Measurements on the Off-Plant Operable Unit**	The Quality Assurance Project Plan (QAPP) and Final Trip Report for the In-Situ Gamma Study perform within the Off-Plant Operable Unit of the Eastern Michaud Flats EPA Superfund Site located near Pocatello, Idaho. ESFs are recommending these documents as examples for sampling/monitoring planning purpose.	electronic file upon request
11.	United States Environmental Protection Agency (US EPA). Multi- **Agency Radiation Survey and Site Investigation Manual (MARSSIM) Training**	The Multi-Agency Radiation Survey and Site Investigation Manual (MARSSIM) provides detailed guidance for planning, implementing, and evaluating environmental and facility radiological surveys conducted to demonstrate compliance with a dose-or risk-based regulation. The MARSSIM guidance focuses on the demonstration of compliance during the final status survey following scoping, characterization, and any necessary remedial actions	http://www.trainex.org/classdetails.cfm?courseid=292&classid=5826
12.	United States Environmental Protection Agency (US EPA). **Multi-Agency Radiation Survey and Site Investigation Manual (MARSSIM)**	The Multi-Agency Radiation Survey and Site Investigation Manual (MARSSIM) provides guidance to federal agencies, states, site owners, contractors, and other private entities on how to demonstrate that their site is in compliance with a radiation dose or risk-based regulation, otherwise known as a release criterion.	http://www.epa.gov/rpdweb00/marssim/
13.	United States Environmental Protection Agency (US EPA). **Multi-Agency Radiological Laboratory Analytical Protocols Manual**	MARLAP provides guidance for the planning, implementation, and assessment phases of those projects that require the laboratory analysis of radionuclides. The guidance provided by MARLAP is both scientifically rigorous and flexible enough to be applied to a diversity of projects and programs. This guidance is intended for project planners, managers, and laboratory personnel.	http://www.epa.gov/rpdweb00/marlap/

(Continued)

Environmental monitoring			
Material		**Description**	**Link**
14.	United States Environmental Protection Agency (US EPA). Soil Screening Guidance for Radionuclides	The Soil Screening Guidance for Radionuclides is a tool developed by EPA to help standardize and accelerate the evaluation and cleanup of radioactively contaminated soils at sites on the National Priorities List (NPL) where future residential land use is anticipated. The User's Guide provides a simple step-by-step methodology for environmental science/engineering professionals to calculate risk-based, site-specific soil screening levels (SSLs) for radionuclides in soil that may be used to identify areas needing further investigation at NPL sites. The risk assessment procedures have been superseded by the PRG calculator, but the radiation survey approach has not.	http://www.epa.gov/superfund/health/contaminants/radiation/radssg. htm
15.	University of Tennessee/Oak Ridge National Laboratory. Spatial Analysis and Decision Assistance	Spatial Analysis and Decision Assistance (SADA) is free software that incorporates tools from environmental assessment fields into an effective problem solving environment. These tools include integrated modules for visualization, geospatial analysis, statistical analysis, human health risk assessment, ecological risk assessment, cost/benefit analysis, sampling design, and decision analysis.	http://www.sadaproject.net/index.html
16.	Whicker, F. W., J. E. Pinder III, J. W. Bowling, J. J. Alberts, and I. L. Bisbin, jr. **"Distribution of Long-lived Radionuclides in an Abandoned Reactor Cooling Reservoir."** Ecological Monographs. Vol. 60, N0. 4, Dec. 1990, pp. 471-496.	The distribution of ^{137}Cs, 90Sr, 238Pu, 239, 240Pu, 241Am and 244Cm was studied in the biotic and abiotic components of an abandoned reactor cooling impoundment, Pond B. The impoundment is located at the United States Department of Energy's Savannah River Plant in South Carolina, USA. It received radioactive contaminants via cooling water discharges from Reactor from September 1961 to June 1964. The radionuclide inventories were estimated in water, seston, sediments, and biotic components after 20 yr of equilibration. Chemical, physical, and biological relationships to the radionuclide distribution patterns were investigated. Biotic components contained some of the highest radionuclide concentration ratios observed to date. However, most of the radioactivity resides in sediments. The principal mechanisms of loss from the system are radioactive decay and periodic outflow of water and suspended materials; biotic export and	http://www.jstor.org/discover/10.2307/1943017?uid=3739896&uid=2134&uid=2&uid=70&uid=4&uid=3739256&sid=21101790738353

Environmental monitoring			
	Material	**Description**	**Link**
		seepage appear to be inconsequential. Strontium—90 was much more mobile in the system than the other radionuclides. Aquatic macrophytes dominated the biotic component radionuclide inventories and their dynamics exert a strong influence on the spatial distribution and turnover of radioactivity in the ecosystem. Pond B supports a diverse and productive flora and fauna. Cleanup of the system is not indicated. Use of Pond B for recreation is feasible with adequate attention to monitoring and radiological health guidelines.	
Cesium Behavior in the Environment			
1.	Beals, D. M., K. J. Hofstetter, L. S. Nichols. ***Cesium-^{137} in the Sediments of Fourmile Creek (U)***. WSRC-TR-2002-00253. Savannah River Technology Center. 2002.	The Nonproliferation Technology Section (NTS) was requested by the Environmental Restoration Division (ER) to aid in completing ground-truth measurements of aerial overflight data in support of the Integrator Operable Unit (IOU) program at the Savannah River Site (SRS). The IOU's at the SRS are under investigation as a possible pathway for the release of contamination from past SRS activities to off-unit receptors and the environment. The IOU's are defined as surface water bodies and associated wetlands, including the water, sediment and related biota. The objective of the IOU program is to: assess the risk to potential human and ecological receptors from IOU contamination; evaluate the impact of inactive and active waste units and operating facilities on the IOU quality; determine if IOU early actions, including reprioritization of operable units implementation schedules, are necessary; and complete the remedial investigation/feasibility study process, defining the nature and extent of IOU contamination, remedial action objectives, and final remediation goals.	http://www.osti.gov/bridge/product.biblio.jsp?query_id=0&page=0&osti_id=805606&Row=0&formname=basicsearc h.jsp
2.	Hinton, T. G., D. I. Kaplan, A. S. Knox, D. P. Coughlin, R. V. Nascimento, S. I. Watson, D. E. Fletcher, B. J. Koo (Savannah River Ecology Laboratory and Savannah River National Laboratory). **"Use of Ilite Clay for In**	We hypothesized that adding micaceous minerals to ^{137}Cs-contaminated aquatic systems would serve as an effective in situ remediation technique by sequestering the contaminant and reducing its bioavailability. Results from several laboratory studies are presented from which an effective amendment material was chosen for a replicated field study. The field study was conducted over a 2-year	http://pubs.acs.org/doi/pdf/10.1021/es060124 x

(Continued)

Cesium Behavior in the Environment			
Material		Description	Link
	Situ Remediation of ^{137}Cs-Contaminated Water Bodies: Field Demonstration of Reduced Biological Uptake." *Environmental Science and Technology. Volume 40, Issue 14, July 15, 2006, Pages 4326-4528.*	period and incorporated 16 3.3-m diameter columnplots (limnocorrals) that were randomly placed in a ^{137}Cs-contaminated pond. The limnocorrals received three rates of amendment treatments to their water surfaces. The amendment material was a commercially available mineral with high sorption (Kd > 9000 L kg-1) and low desorption (<20%) characteristics for cesium, even in the presence of high concentrations of the competing cation, NH4 +. In the treated limnocorrals, ^{137}Cs concentrations were reduced some 25-30-fold in the water, 4-5-fold in aquatic plants, and 2-3-fold in fish. The addition of the amendment did not adversely affect water chemistry, although increased turbidity and subsequent siltation did alter the aquatic macroinvertebrate insect community. This in situ technology provides a valuable, less-environmentally intrusive alternative to costly ex situ technologies that require the contaminated sediment to be excavated prior to treatment, or excavated and disposed of elsewhere.	
3.	Johnson BE, PH Santschi, RS Addleman, M Douglas, JD Davidson, GE Fryxell, and JM Schwantes. (2011). "Collection of fission and activation product elements from fresh and ocean waters: a comparison of traditional and novel sorbents." Applied Radiation and Isotopes 69(1):205-216. doi:10.1016/j.apradiso.2010.07.025	Monitoring natural waters for the inadvertent release of radioactive fission products produced as a result of nuclear power generation downstream from these facilities is essential for maintaining water quality. To this end, we evaluated sorbents for simultaneous in-situ large volume extraction of radionuclides with both soft (e.g., Ag) and hard metal (e.g., Co, Zr, Nb, Ba, and Cs) or anionic (e.g., Ru, Te, Sb) character. In this study, we evaluated a number of conventional and novel nanoporous sorbents in both fresh and salt waters. In most cases, the nanoporous sorbents demonstrated enhanced retention of analytes. Salinity had significant effects upon sorbent performance and was most significant for hard cations, specifically Cs and Ba. The presence of natural organic matter had little effect on the ability of chemisorbents to extract target elements.	http://www.sciencedirect.com/science/journal/09698043/69/1

Cesium Behavior in the Environment			
Material		Description	Link
4.	Onishi, Y., O.V. Voitsehkovich, and M.J. Zheleznyak, edts. 2007. Chernobyl – What Have We Learned? The Successes and Failures to Mitigate Water Contamination Over 20 Years, Springer Publishers, Dordrecht, The Netherlands.	Twenty million people have been exposed to Chernobyl radionuclides through the Dnieper River aquatic pathways. This book presents a 20-year historical overview and comprehensive study results of the aquatic environment affected by the 1986 Chernobyl nuclear accident. During this time, many water quality management practices and countermeasures were enacted. The book presents in-depth analyses of these water remediation actions, using current science and mathematical modeling, and discusses why some were successful, but many others failed.	http://www.springer.com/environment/book/978-1-4020-5348-1
5.	Peters, E. L., and I. L. Bisbin, jr. "Environmental Influences on the ^{137}Cs Kinetics of the Yellow-Bellied Turtle (Trachemys Scripta)." Ecological Monographs. Vol. 66, No. 1, Feb. 1996. pp. 113-136.	Assessments of ecological risk require accurate predictions of contaminant dynamics in natural populations. However, simple deterministic models that assume constant uptake rates and elimination fractions may compromise both their ecological realism and their general application to animals with variable metabolisms or diets. In particular, the temperature- dependent metabolic rates characteristic of ectotherms may lead to significant differences between observed and predicted contaminant kinetics. We examined the influence of a seasonally variable thermal environment on predicting the uptake and annual cycling of contaminants by ectotherms, using a temperature-dependent model of ^{137}Cs kinetics in free- living yellow-bellied turtles, Trachemys scripta. We compared predictions from this model with those of deterministic negative exponential and flexibly shaped Richards sigmoidal models. Concentrations of ^{137}Cs in a population of this species in Pond B, a radionuclide-contaminated nuclear reactor cooling reservoir, and ^{137}Cs uptake by uncontaminated turtles held captive in Pond B for 4 yr confirmed both the pattern of uptake and the equilibrium concentrations predicted by the temperature-dependent model. Almost 90% of the variance in the predicted time-integrated ^{137}Cs concentration was explainable by linear relationships with model parameters. The model was also relatively insensitive to uncertainties in the estimates of ambient temperature, suggesting that adequate estimates of temperature- dependent ingestion and elimination may	http://www.jstor.org/discover/10.2307/2963483?uid=3739896&uid=2134&uid=2&uid=70&uid=4&uid=3739256&sid=21101790738353

(Continued)

Cesium Behavior in the Environment			
Material		Description	Link
		require relatively few measurements of ambient conditions at sites of interest. Analyses of Richards sigmoidal models of ^{137}Cs uptake indicated significant differences from a negative exponential trajectory in the 1st yr after the turtles' release into Pond B. We also observed significant annual cycling of ^{137}Cs concentrations, apparently due to temperature-dependent metabolism and its influence on ingestion and elimination rates. However, equilibrium concentrations of the radionuclide in the wild population were predictable from negative exponential models based on average annual temperature and its effects on intake and elimination rates, also suggesting that predicting ectotherm responses to long-lived contaminants (such as ^{137}Cs) may be possible without complex ecophysiological modeling.	
6.	Serkiz, S. (Westinghouse Savannah River Company). **"An in situ method for remediating ^{137}Cs-contaminated wetlands using naturally occurring minerals."** *Journal of Radioanalytical and Nuclear Chemistry.* Vol. 249, No. 1, July 2001, pp. 197-202)	Cesium's enhanced bioavailability in contaminated wetlands on the U.S. Department of Energy's Savannah River Site (SRS) is thought to be due to the low clay fraction of SRS soils, and that the clay mineralogy is dominated by kaolinites. Remediation of the wetlands is problematic because current technologies are destructive to the sensitive ecosystems. We tested11 clay minerals (two micas, a vermiculite, six illites, a kaolinite, and a smectite) for their propensity to sorb and retain ^{137}Cs. Two minerals were subsequently chosen as candidates for in situ remediation amendment materials because they had ^{137}Cs distribution coefficients (K d) well in excess of 10,000 ml . g –1 , and desorbed less than 20% of the Cs when mixed in a 0.1M NH 4 Cl solution. Incremental additions of the candidate minerals to ^{137}Cs-contaminated sediments appreciably intercepted and retained desorbed ^{137}Cs in the presence of high levels of NH4 . Implications for using the minerals as a nondestructive, in situ remediation technique are discussed.	http://link.springer.com/article/10.1023/A%3A1013234309124

Cesium Behavior in the Environment			
	Material	**Description**	**Link**
7.	National Oceanic and Atmospheric Administration (NOAA). ***Anthropogenic Radionuclides in the Marine Environment: A Selected Bibliography***.	In response to the ongoing situation at Japan's Fukushima Daiichi nuclear power plant, librarians at the NOAA Central Library, in collaboration with NOAA's Office of Oceanic and Atmospheric Research and National Marine Fisheries Service, have prepared this bibliography on the distribution, transport, ecological effects, and potential risk to human health of anthropogenic radionuclides in the marine environment.	http://www.lib.noaa.gov/researchtools/subjectguides/marine_radionu clides.html#
8.	International Atomic Energy Agency (IAEA). ***Modelling and Data for Radiological Impact Assessments (MODARIA).***	The general aim of the MODARIA Programme is to improve capabilities in the field of environmental radiation dose assessment by means of acquisition of improved data for model testing, model testing and comparison, reaching consensus on modelling philosophies, approaches and parameter values, development of improved methods and exchange of information.	http://www-s.iaea.org/projects/m odaria/default.asp?s=8&l=116
9.	United States Environmental Protection Agency (US EPA). ***Understanding Variation in Partition Coefficient, K_d, Values.***	This three-volume report describes the conceptualization, measurement, and use of the partition (or distribution) coefficient, Kd, parameter. It also describes the geochemical aqueous solution and sorbent properties that are most important in controlling adsorption/retardation behavior of selected contaminants.	http://www.epa.gov/radiation/cleanup/402-r-99-004.html
10.	United States Environmental Protection Agency (US EPA). ***Fate of Radiological Dispersal Device (RDD) Material on Urban Surfaces: Impact of Rain on Removal of Cesium.***	This study investigated the effect of rain on the fate of cesium (Cs) on urban surfaces. The rinsed amount of Cs by rain from contaminated surfaces was measured and the distribution of Cs at and below the surface was characterized. Results from the study on fate of cobalt (Co), conducted in parallel with this study, will be part of a separate report.	http://cfpub.epa.gov/si/si_public_file_download.cfm?p_download_id=507700
11.	United States Environmental Protection Agency (US EPA). ***Assessment of the Fate of RDD Contamination after Laundering of Soft Porous Materials.***	This project was designed to help develop decontamination efficacy data related to the laundering of clothing and other porous soft materials contaminated due to a radiological dispersal device (RDD).	http://cfpub.epa.gov/si/si_public_file_download.cfm?p_download_id=508206
Remediation Strategy			
1	US DOE Hanford Site Cleanup Completion Framework	Maintains an overall strategy for organizing completion of the multi-decade large-scale remediation of DOE's Hanford Site in southeastern Washington state. Organizes the completion strategy into three major	http://www.hanford.gov/page.cfm/HanfordSiteCleanupCompletionFramewor k

(Continued)

	Material	Description	Link
Remediation Strategy			
		components: River Corridor, Central Plateau, and Tank Waste. Provides an overall framework for showing how all major parts of the remediation strategy support completion of the cleanup mission.	
2	US DOE 2013 Hanford Lifecycle Scope, Schedule and Cost Report (Lifecycle Report)	The Lifecycle Report reflects all cleanup work that must be completed to fully meet all applicable environmental obligations. Provide project scope, schedule and cost from Fiscal Year (FY) 2013 – FY 2090, and will be modified annually; Includes scope, schedule and cost information across the entire Hanford Site regardless of whether the final cleanup decision has been made;	http://www.hanford.gov/files.cfm/DOE-RL-2012-13_FINAL__REV.0_.pdf
3	US DOE Proposed Plan for the Remediation of the 200-CW-5, 200-PW-1, 200-PW-3, and 200-PW-6 Operable Units	This Proposed Plan is an example of a CERCLA document that proposes a remediation approach for a set of Hanford waste sites contaminated with $Cs\text{-}^{137}$ and plutonium. It is provided for public comment on the proposed remedy prior to formally issuing a record of decision. Public input into these decisions is an essential element of the CERCLA cleanup process.	http://www5.hanford.gov/arpir/?content=findpage&AKey=0093832
4	US DOE Feasibility Study for the Plutonium/Organic-Rich Process Condensate/Process Waste Group Operable Unit: Includes the 200-PW-1, 200-PW-3, and 200-PW-6 Operable Units	This CERCLA Feasibility Study is the companion document to item #3 above. The Feasibility Study evaluates a range of remediation approaches from "no action" to complete removal with off-site disposal. The report evaluates all remediation alternatives for nine CERCLA criteria and provides the technical analysis that supports development of a proposed and final decision.	http://www5.hanford.gov/arpir/?content=findpage&AKey=0093807
Cross-cutting Considerations			
Citizens Advisory Board for Decision-Making on U.S.Environmental Restoration Projects			
1.	Belencan, Helen L, Guevara, Karen C, and Spears, Terrel J. "A Retrospective Management Perspective on Nearly 20 Years of the Savannah River Site Citizen Advisory Board" paper number 13078, WM2013 Conference, February 24 – 28,	The Department of Energy, Office of Environmental Management (DOE EM) program has invested in site specific advisory boards since 1994. These boards have served as a portal to the communities surrounding the DOE sites, provided a key avenue for public involvement, and have actively engaged in providing input and feedback that has informed clean up and priority decisions made by	http://www.srs.gov/general/outreach/srs-cab/srs-cab.html

Cross-cutting Considerations			
Citizens Advisory Board for Decision-Making on U.S.Environmental Restoration Projects			
Material		**Description**	**Link**
	2013, Phoenix, Arizona, USA	EM. Although the EM program has made considerable progress in completing its mission, work will continue for decades, including work at the Savannah River Site (SRS). It is reasonable to assume the advisory boards will continue in their role providing input and feedback to EM. The SRS Citizen Advisory Board (CAB) formed in 1994 and has issued 298 recommendations through September 2012. Although the effectiveness of the board is not measured by the number of recommendations issued, the recommendations themselves serve to illustrate the areas in which the CAB is particularly interested, and offer insight to the overall effectiveness of the CAB as a means for public participation in the EM decision making process.	
2.	Elder, Susan. **The Citizens' Advisory Board, BELLA / SRS Supplement**	BELLA Magazine feature article on the SRS CAB	http://www.srs.gov/ge neral/outreach/srs-cab/srs-cab.html
3.	Savannah River Site Citizens Advisory Board (CAB) Standard Operating Procedures, Approved 12/13/07	The Savannah River Site (SRS) Citizens Advisory Board (CAB) provides the Assistant Secretary for Environmental Management and designees with advice, information, and recommendations on issues affecting the EM program. Among those issues are clean-up standards and environmental restoration; waste management and disposition; stabilization and disposition of non-stockpile nuclear materials; excess facilities; future land use and long-term stewardship; risk assessment and management; and clean-up science and technology activities. The board's membershi9p is carefully considered to reflect a full diversity of viewpoints in the community and region. Board members are composed of people who are directly affected by DOE site clean-up activities.	http://www.srs.gov/general/outreach/srs-cab/srs-cab.html
4.	U. S. Department Of Energy Environmental Management Site-Specific Advisory Board (EM SSAB) Savannah River Site Application Form 8/2/2012	United States Department of Energy Office of Environmental Management. Savannah River Site – Citizens Advisory Board MEMBERSHIP APPLICATION FORM	http://www.srs.gov/general/outreach/srs-cab/srs-cab.html

(Continued)

Community Engagement			
Material		Description	Link
1.	United States Environmental Protection Agency (US EPA)	webpage and flyers for EPA response after World Trade Center incident. Examples of community engagement after a large scale event.	http://www.epa.gov/katrina/index.html http://www.epa.gov/katrina/outreach/hando uts.html http://www.epa.gov/katrina/outreach/returni ng-general.pdf http://www.epa.gov/katrina/outreach/psa.ht ml
2.	United States Environmental Protection Agency (US EPA)	webpage, handouts, and public service announcements for EPA response after Hurricanes Katrina and Rita	http://www.epa.gov/wtc/ http://web.archive.org/web/20100508195620/http://www.epa.gov/wtc/flyers/onepagead. pdf http://web.archive.org/web/20100508195634 /http://www.epa.gov/wtc/flyers/newspaper_ ad.pdf
3.	United States Environmental Protection Agency (US EPA)	webpage, fact sheets for EPA response after Hurricane Sandy	http://www.epa.gov/sandy/index.html http://www.epa.gov/sandy/factsheets.htmlhttp://www.epa.gov/sandy/disinfectdrinkingwater.html http://www.epa.gov/sandy/frequentquestions .html
4.	United States Environmental Protection Agency (United States Environmental Protection Agency (US EPA) Public Involvement Policy	The purposes of this Policy are to improve the acceptability, efficiency, feasibility and durability of the Agency's decisions, reaffirm EPA's commitment to early and meaningful public involvement, ensure that EPA makes its decisions considering the interests and concerns of affected people and entities, promote the use	**http://www.epa.gov/policy2003/policy2003 .htm**

Community Engagement

	Material	Description	Link
		of a wide variety of techniques to create early and, when appropriate, continuing opportunities for public involvement in Agency decisions, and establish clear and effective guidance for conducting public involvement activities.	
5.	United States Environmental Protection Agency (US EPA)	Public Involvement Policy of the US Environmental Protection Agency	http://www.epa.gov/publicinvolvement/policy2003/finalpolicy.pdf
6.	United States Environmental Protection Agency (US EPA)	EPA Superfund Community Involvement	http://www.epa.gov/superfund/community/index.h tm
7.	United States Environmental Protection Agency (US EPA)	Lessons learned about EPA Superfund Community Involvement, EPA Superfund Response Staff Tell How Public Involvement Has Helped Clean Up Sites	http://www.epa.gov/superfund/programs/reforms/docs/leslrncomplete.pdf
8.	United States Environmental Protection Agency (US EPA) Better Decisions through Consultation and Collaboration, Conflict Prevention, and Resolution Center	This document is a resource guide on public involvement best practices and strategies for EPA staff who are tasked with designing and/or implementing public involvement processes for various EPA activities. The discussions and advice in this document are intended solely as guidance.	http://www.epa.gov/adr/Better_Decisions.pd f
9.	United States Environmental Protection Agency (US EPA) community involvement	Community involvement interviews	http://www.epa.gov/region5/cleanup/kerrmcgee/200807.htm
10.	United States Environmental Protection Agency (US EPA) **Framework for Implementing United States Environmental Protection Agency (US EPA)'s Public Involvement Policy**	This document outlines activities for three critical internal functions: training, information sharing and evaluation.	http://www.epa.gov/publicinvolvement/policy2003/framework.pdf
11.	United States Environmental Protection Agency (US EPA) Public meeting notice	public meeting notification example	http://www.epa.gov/reg3hwmd/npl/PAD987295276/pn/SLC_ROD_public_notice_091710.pdf

(Continued)

Community Engagement			
Material		Description	Link
12.	United States Environmental Protection Agency (US EPA) public meeting invitation	Public meeting invitation example	http://www.epa.gov/region5/cleanup/kerrmcgee/pdfs/kerrmcgee_pc_201104.pdf
13.	United States Environmental Protection Agency (US EPA) public meeting proceedings	public meeting proceedings	http://www.epa.gov/region5/sites/ottawa/pdfs/ottawa_hearing_201 005.pdf
14.	United States Environmental Protection Agency (US EPA) Stakeholder involvement and public participation at the U.S. EPA	In the 1990s, EPA increased its efforts to involve the public by giving citizens, industry, environmental groups, and academics a much greater opportunity to play key roles in environmental decision-making. Today, EPA is continuing this tradition by initiating and supporting a vast array of stakeholder involvement and public participation initiatives well beyond the scope of what was originally in place when the Agency opened for business in 1970. Due to the diversity and extensive number of Agency initiatives involving the public, however, much of the wisdom and experience gained by EPA staff implementing these efforts can be lost from one activity to the next, making it difficult for the rest of the Agency to benefit. Staff performing outreach and leading stakeholder involvement and public participation activities in one office may have limited interaction with staff performing similar types of work in other offices. In addition, Agency reviews of stakeholder involvement and public participation tend to focus on single initiatives and preclude Agency staff from benefitting from a broader perspective of EPA's public involvement activities.	http://www.epa.gov/evaluate/pdf/stakeholder/stakeholder-involvement-public-participation-at-epa.pdf
15.	United States Environmental Protection Agency (US EPA) The Model Plan for Public Participation	This report and recommendations have been written as a part of the activities of the National Environmental Justice Advisory Council (NEJAC), a public advisory committee providing extramural policy information and advice to the Administrator and other officials of the United States Environmental Protection Agency (EPA). The Council	http://www.epa.gov/compliance/ej/resources/publications/nejac/model-public-part-plan.pdf

Community Engagement			
Material		Description	Link
		is structured to provide balanced, expert assessment of matters related to environmental justice. This report has been reviewed by the EPA. Mention of trade names or commercial products does not constitute a recommendation for use.	
16.	United States Environmental Protection Agency (US EPA) tool	How to Participate in EPA Decision-Making (63 FR 58038, October, 1998) Resource Guide for Resolving Conflicts in Communities (EPA/360-F-00-001, May 2001)	http://www.epa.gov/a dr/Resguide.pdf
17.	United States Environmental Protection Agency (US EPA) tool	Guide on Consultation and Collaboration with Indian Tribal Governments and the Public Participation of Indigenous Groups and Tribal Citizens, National Environmental Justice Advisory Committee, Indigenous Peoples Subcommittee, November, 2000	http://www.epa.gov/compliance /resources/publications/ej/ips consultation guide.pdf
18.	United States Environmental Protection Agency (US EPA) tool	Model Plan for Public Participation (EPA/300-K-96-003, November 1996)	http://www.epa.gov/compliance /resources/publications/ej/mode l_p ublic_part_plan.pdf
19.	United States Environmental Protection Agency (US EPA) tool	Environmental Justice in EPA Permitting: Reducing Risk in High-Risk Communities is Integral with the Agency's Mission, National Academy of Public Administration, December, 2001	http://www.epa.gov/c ompliance/resources/p ublications/ej/napa-epa-permitti ng.pdf
20.	United States Environmental Protection Agency (US EPA) tool	Community Culture and the Environment: A Guide to Understanding a Sense of Place (EPA 842-B-01-003, January 2003)	http://www.epa.gov/e cocommunity/pdf/ccec omplete.pdf
21.	United States Environmental Protection Agency (US EPA) tool	RCRA Public Involvement Manual (EPA530-R-96-007, September 1996) http://www.epa.gov/epaoswer/hazwaste/permit/pubpart/manual.htm	http://www.epa.gov/e paoswer/hazwaste/per mit/pubpart/manual.ht m
22.	United States Environmental Protection Agency (US EPA) tool	RCRA Public Involvement Manual (EPA 530-R-96-007, September 1996)	http://www.epa.gov/e paoswer/hazwaste/per mit/pubpart/manual.ht m
23.	United States Environmental Protection Agency (US EPA) tool	Information Products Bulletin	http://www.epa.gov/ip bpages
24.	United States Environmental Protection Agency (US EPA) tool	Libraries at EPA	http://www.epa.gov/n atlibra/libraries.htm

(Continued)

Community Engagement			
Material		Description	Link
25.	United States Environmental Protection Agency (US EPA) tool	Office of Cooperative Environmental Management	http://www.epa.gov/o cem
26.	United States Environmental Protection Agency (US EPA) tool	Strategic Information Plan: A Framework for the Future (July 29, 2002)	http://www.epa.gov/o ei/collecting.htm
27.	United States Environmental Protection Agency (US EPA) tool	Regulations and Proposed Rules Web site	http://www.epa.gov/o pei/regulatory.htm
28.	United States Environmental Protection Agency (US EPA) tool	Public Involvement in Environmental Permits: A Reference Guide (EPA599-R00-007, August 2000)	http://www.epa.gov/p ermits/publicguide.htm
29.	United States Environmental Protection Agency (US EPA) tool	Public Involvement Web site	http://www.epa.gov/p ublicinvolvement
30.	United States Environmental Protection Agency (US EPA) tool	Small Business Gateway Web site: see http://www.epa.gov/smallbusiness/	http://www.epa.gov/s mallbusiness/
31.	United States Environmental Protection Agency (US EPA) tool	Public Involvement Techniques for Transportation Decision-making, US Department of Transportation, Federal Highway Administration, Federal Transit Administration (FHWA-PD-96-031HEP-30/9-96/(4M)QE, September 1996)	http://www.fhwa.dot.g ov//////reports/pittd/c ontents.htm
32.	United States Environmental Protection Agency (US EPA) tool	Social Aspects of Siting RCRA Hazardous Waste Facilities (EPA/530–K-00-005, April, 2000)	www.epa.gov/epaoswe r/hazwaste/tsds/site/k 00005.pdf
33.	United States Environmental Protection Agency (US EPA) training	The U.S. Environmental Protection Agency's (EPA) 13th Community Involvement Training Conference will be held in Boston, Massachusetts July 30-August 1, 2013. The EPA Office of Water, EPA Region 1, and the EPA Office of Solid Waste and Emergency Response are leading the planning efforts for this conference.	http://www.epa.gov/ci conference/
Example fact sheets to distribute to the stakeholders			
1	United States Environmental Protection Agency (US EPA) fact sheet	Fact sheet on Proposed Plan to clean up Groundwater OU in Central Part of Hanford (July 2012)	http://www2.hanford.g ov/arpir/?content=find page&AKey=0092338

Example fact sheets to distribute to the stakeholders			
Material		Description	Link
2	United States Environmental Protection Agency (US EPA) fact sheet	Fact sheet on Record of Decision to clean up cesium and plutonium soil sites at Hanford	http://www2.hanford.gov/arpir/?content=findpage&AKey=0093618
3	United States Environmental Protection Agency (US EPA) fact sheet	Fact sheet on Proposed Plan to clean up cesium and plutonium soil sites at Hanford (July 2011)	http://www2.hanford.gov/arpir/?content=findpage&AKey=0093833
4	United States Environmental Protection Agency (US EPA) fact sheet	Fact sheet on Clean-up Schedule Announced for Contaminated River and Creek	http://www.epa.gov/region5/cleanup/kerrmcgee/pdfs/kerrmcgee_fs_200508.pdf
5	United States Environmental Protection Agency (US EPA) fact sheet	Fact sheet on EPA proposed changes to plan for vacant properties cleanup	http://www.epa.gov/region5/cleanup/ottawa/index.htm#factsheets
6	United States Environmental Protection Agency (US EPA) fact sheet	EPA Facts About Cesium-137	http://www.epa.gov/superfund/health/contaminants/radiation/pdfs/cesium.pdf
Example videos to encourage the public involvement			
2	United States Department of Energy (US DOE)/United States Environmental Protection Agency (US EPA)/ Washington State Department of Ecology video	a video to encourage the public to attend workshops on Hanford cleanup along the Columbia River (June 2012)	http://www.youtube.com/watch?v=yqEh13aNGbU&list=UUIks5sbjPPyDevywkmbGSSQ&index=11
3	United States Department of Energy (US DOE)/United States Environmental Protection Agency (US EPA)/ Washington State Department of Ecology video	a video for the Hanford Public Involvement Plan to encourage the public to provide input on proposed changes to the document. (Oct 2011)	http://www.youtube.com/watch?v=8HrRddxXu64&list=UUIks5sbjPPyDevywkmbGSSQ&index =21
1	United States Environmental Protection Agency (US EPA) education video	EPA Superfund Radiation Risk Assessment and How You Can Help: An Overview: 19 minute video on rad risk assessment for the general public	http://www.epa.gov/superfund/health/contaminants/radiation/radv ideo.htm

(Continued)

Example videos to encourage the public involvement			
Material		Description	Link
4	United States Environmental Protection Agency (US EPA) video	EPA and Navajo Nation EPA have worked together to identify all the structures that should be assessed. Together we have assessed a total of 763 structures to date and continue to refer structures for more assessment and remediation. Residents will be notified of the results and possible remedial solutions if any are warranted. EPA Superfund authority will be used to perform any remediation.	http://www.epa.gov/region9/superfund/Navajonation/contaminated-structures.html

End Notes

[1] Monitored Natural Attenuation - "Reliance on natural attenuation processes (within the context of a carefully controlled and monitored site cleanup approach) to achieve site-specific remediation objectives within a time frame that is reasonable compared to that offered by other more active methods. The 'natural attenuation processes' that are at work in such a remediation approach include a variety of physical, chemical, or biological processes that, under favorable conditions, act without human intervention to reduce the mass, toxicity, mobility, volume, or concentration of contaminants in soil or groundwater. These in-situ processes include biodegradation; dispersion; dilution; sorption; volatilization; radioactive decay; and chemical or biological stabilization, transformation, or destruction of contaminants." - U.S. Environmental Protection Agency, 1999

[2] http://www.jaea.go.jp/fukushima/kankyoanzen/guideline_report.html; http://www.jaea.go.jp/fukushima/kankyoanzen/d-model_report.html

[3] http://if.quintessa.co.jp/c-navi/en/.

[4] http://josen.env.go.jp/en/

[5] http://josen.env.go.jp/en/pdf/annex_03.pdf

[6] "Decontamination Roadmap" for the Special Decontamination Area, Released by MOE on January 26, 2012.

[7] http://www.env.go.jp/en/focus/docs/files/20120917-46.pdf

[8] Briefing material for technical workshop on the accident at Fukushima Dai-ichi nuclear power station, TEPCO by Cabinet Office, Japan, Support Team for Residents Affected by Nuclear Incidents - See more at: http://www.meti.go.jp/english/earthquake/nuclear/roadmap/pdf/20120723_01.pdf

[9] The Long-term Strategy for Remediation, Ministry of the Environment, January 28th, 2013, International Experts' meeting on Decommissioning and Remediation after a Nuclear Accident, Vienna, Austria, http://wwwpub.iaea.org/iaeameetings/IEM4/Session1/Nishiyama.pdf

[10] Progress on Offsite Cleanup Efforts in Japan, Ministry of the Environment, Japan at Senior Regulators' Meeting at 56th IAEA General Conference, September 20th, 2012

[11] Booklet prepared by Masaru Moriya, Head of MOE's Fukushima Decontamination Team and the Fukushima Prefectural Government (in Japanese) during the meeting on February 22nd

[12] Yasumura et al., Study Protocol for the Fukushima Health Management Survey, J. epidemiol. 2012; 22(5); 375-383. http://www.fmu.ac.jp/radiationhealth/publications/media/Yasumura_S_et_al_J_Epidemiol.pdf

[13] Information about the Fukushima Health Management Survey, http://www.fmu.ac.jp/univ/chiiki/health_survey/pdf/en/en_zip.pdf

[14] Proceedings of the 8th Committee Meeting for Fukushima Health management Survey, September 11th 2012. http://www.fmu.ac.jp/radiationhealth/results/20120911.html

[15] Proceedings of the 10th Committee Meeting for Fukushima Health management Survey, February 13th 2013. http://www.fmu.ac.jp/radiationhealth/results/10-1_basic_survey.html

[16] 復興庁「避難住民説明会等でよく出る放射線リスクに関する質問・回答集（平成24年12 月25 日）」, pp.12 (http://www.reconstruction.go.jp/topics/20121225_risukomisiryour1.pdf).

[17] Nuclear Regulation Authority of Japan, http://www.nsr.go.jp/archive/nsc/mext_speedi/

[18] Kim et al., Screening Survey on Thyroid Exposure for Children after the Fukushima Daiichi Nuclear Power Station Accident, *The 1st NIRS Symposium on Reconstruction of Early Internal Dose in the TEPCO Fukushima Daiichi Nuclear Power Station Accident*, pp.59-66 (http://www.nirs.go.jp/publication/irregular/04.shtml).

[19] Application of the Commission's Recommendations to the Protection of People Living in Long-term Contaminated Areas After a Nuclear Accident or a Radiation Emergency, ICRP Publication 111, 2009

[20] John Cardarelli II, Larry Elliott, Richard Hornung, Wushou P. Chang Health Phys. 72, No. 3, pp. 351-360 (March 1997).

[21] Health risk assessment, World Health Organization, 2013,http://apps.who.int/iris/bitstream/10665/78218/1/9789241505130_eng.pdf

[22] External Dosimetry Technical Basis Manual , WSRC-IM-92-101, Revision 16, January 16, 2013, Savannah River Site.

[23] The Savannah River Site Internal Dosimetry Technical Basis Manual (U), WSRC-IM-90-139, Revision 13, October 25, 2012, Savannah River Site.

[24] http://www.jaea.go.jp/fukushima/kankyoanzen/guideline_report.html; http://www.jaea.go.jp/fukushima/kankyoanzen/d-model_report.html

[25] FY2011 "Decontamination Technology Demonstration Test Project," Yoshitake Shiratori and Akihiro Tagawa, Japan Atomic Energy Agency, presentation at Meeting for Reporting the Results of the Decontamination Demonstration Model Project, March 26, 2012, Fukushima City Public Hall under the auspices of the Cabinet Office's Team in Charge of Assisting the Lives of Disaster Victims, the Ministry of the Environment, and the Japan Atomic Energy Agency.

[26] "Decontamination Techniques Demonstration Program 2011," August 2012, Environment Management Bureau, Ministry of the Environment.

[27] Decontamination Guidelines, 1st edition, December 2011, (tentative English translation), published by the Ministry of the Environment

[28] Common Specifications

[29] "The National Government and Municipalities have Evaluated Decontamination Effectiveness," January 2013, report by the Ministry of the Environment. [Information available on the MOE web site].

[30] Quantifying the amount of decontamination waste that has been generated and its location proved to be difficult. MOE is currently in the process of surveying local communities to collect this information.

[31] Specified waste includes materials that have been contaminated by the accident including debris, sewage sludge, agricultural waste and other items. Ash from incineration of combustible materials can also be categorized as specified waste.

[32] Act on Special Measures concerning the Handling of Environment Pollution by Radioactive Materials Discharged by the NPS Accident Associated with the Tohoku District- Off the Pacific Ocean Earthquake That Occurred on March 11 , 2011, 11/11/2011

[33] ESFs did not study the waste management processes in other prefectures. Per the Basic Policy on Interim Storage (2011), because amounts of removed soil, etc. and designated waste generated in other prefectures are likely to be relatively small and the levels of contamination these are estimated to be relatively low, such soil and waste shall be disposed of in the respective areas within the prefectures through the use of existing controlled landfill sites."

[34] Ministry of the Environment, Decontamination Guidelines, 1st ed., 12/2011 (Tentative Translation)

[35] "Status of Decontamination Storage for Fukushima Prefecture," 12/31/2012 (Japanese only).

[36] Ministry of the Environment, Basic Policy on Interim Storage and Other Facilities Required for the Handling of the Environmental Pollution from Radioactive Materials Associated with the Accident at Tokyo Electric Power Co.'s Fukushima Daiichi Nuclear Power Stations, 10/29/2011

[37] A point of reference is the Environmental Restoration Disposal Facility (ERDF) at the Hanford site in Washington State. There are currently more than 14 million tons of soil and debris occupying a volume of 6.5 million m^3. The project ISF volume is more than four times ERDF's current capacity. Hanford has been filling this disposal site since 1996 and more

than 12,000,000 miles of truck miles have been logged. Source: http://www.hanford.gov/page.cfm/ERDF.

[38] Source: http://www.nsr.go.jp/archive/nisa/shingikai/700/14/240724/AT-6-1.pdf

[39] Source: http://www.nsr.go.jp/archive/nisa/shingikai/700/14/240724/AT-6-1.pdf

[40] Decided on August 2, 2011; Revised on March 15, 2012; Revised on April 1, 2012 http://radioactivity.nsr.go.jp/en/contents/5000/4514/24/204_1_120410e.pdf. Another update as of April 1, 2013 is available in Japanese on the NRA website at http://radioactivity.nsr.go.jp/ja/contents/8000/7302/24/204_moni0401%20(1).pdf

[41] 文部科学省が実施する放射線モニタリング業務の 原子力規制委員会への移管について (Radiation Monitoring Mission Transfer from MEXT to NRA, 文部科学省 原子力災害対策支援本部(MEXT/Nuclear Emergency Response Headquarter) March 22nd 2013, http://radioactivity.nsr.go.jp/ja/contents/7000/6795/24/203_0322_18.pdf

[42] http://radioactivity.nsr.go.jp/en/contents/5000/4513/24/1001_031614e.pdf

[43] http://radioactivity.nsr.go.jp/en/contents/5000/4513/24/1001_031614e.pdf

[44]

- "Sea Area Monitoring Plan in FY12" March 30, 2012 –
- http://radioactivity.nsr.go.jp/en/contents/5000/4854/24/191_e_120330_16.pdf
- "Radiation Monitoring Action Plan for Supporting the Return to/Restoration of Areas to which Evacuation Orders Are Ready to be Lifted (Minami Soma city, Tamura city, Kawauchi village)" (April 18, 2012) – http://radioactivity.nsr.go.jp/en/contents/6000/5604/24/191_120418_17.pdf
- "Radiation Monitoring Action Plan for Supporting Restoration of the Former Emergency Evacuation Preparation Areas (Minami Soma city, Tamura city, Kawauchi village, Hirono town, and Naraha town)" (October 3, 2011) – http://radioactivity.nsr.go.jp/en/contents/1000/338/24/1000_1003_2.pdf
- "Plan to Conduct Detailed Monitoring in Restricted Area and Planned Evacuation Zone" (June 13, 2011) – http://radioactivity.nsr.go.jp/en/contents/1000/317/24/1304084_0613.pdf
- A comprehensive set of monitoring plans and related results (in pdf format) are available at http://radioactivity.nsr.go.jp/en/list/191/list-1.html

[45] http://radioactivity.nsr.go.jp/ja/

[46] http://fukushima-radioactivity.jp

[47] http://radioactivity.nsr.go.jp/ja/

[48] PNNL 2012, PNNL-21379

[49] The present understanding that radioactive cesium is bound in the first several centimeters of the soil is based on findings of NIES. The model G-CIEMS, originally developed by NIES to evaluate fate of chemical substances was applied to evaluate the fate of cesium.

[50] Carlton, W. H. ; Bauer, L. R., Evans, A. G.; Geary, L. A.; Murphy, C. E. Jr.; Pinder, J. E., Strom, R. N. (Westinghouse Savannah River Company). **"Cesium in the Savannah River Site environment."** WSRC-RP-92-250. March 1, 1992.

[51] Whicker, F. W., J. E. Pinder III, J. W. Bowling, J. J. Alberts, and I. L. Bisbin, jr. **"Distribution of Long-lived Radionuclides in an Abandoned Reactor Cooling Reservoir."** Ecological Monographs. Vol. 60, N0. 4, Dec. 1990, pp. 471-496.

[52] Mohler, H. J., F. W. Whicker, T. G. Hinton. **"Temporal trends of ^{137}Cs in an abandoned reactor cooling reservoir."** *Journal of Environmental Radioactivity*. Vol. 37, No. 3, 1997, pp. 251-268.

[53] Abraham, J. P., F. W. Whicker, T. G. Hinton, D. J. Rowan. **"Inventory and spatial pattern of 137Cs in a pond: a comparison of two survey methods."** *Journal of Environmental Radioactivity*. Vol. 51, No. 2, 2000, pp. 157-171.

[54] Beals, D. M., K. J. Hofstetter, L. S. Nichols. ***Cesium-^{137} in the Sediments of Fourmile Creek (U)***. WSRC-TR-2002-00253. Savannah River Technology Center. 2002.

[55] Dion, H. M., C. S. Romanek, T. G. Hinton, P. M. Bertsch. **"Cesium-^{137} in floodplain sediments of the Lower Three Runs Creek on the DOE Savannah River Site."** *Journal of Radioanalytical and Nuclear Chemistry*. Vol. 264, No. 2, 2005, pp. 481-488.

[56] Peters, E. L., and I. L. Bisbin, jr. **"Environmental Influences on the ^{137}Cs Kinetics of the Yellow-Bellied Turtle (Trachemys Scripta)."** *Ecological Monographs*. Vol. 66, No. 1, Feb. 1996. pp. 113-136.

[57] Hinton, T. G., A. Knox, D. Kaplan, S. Serkiz. **"An in situ method for remediating ^{137}Cs-contaminated wetlands using naturally occurring minerals."** *Journal of Radioanalytical and Nuclear Chemistry*. Vol. 249, No. 1, 2001, pp. 197-202.

[58] Hinton, T. G., D. I. Kaplan, A. S. Knox, D. P. Coughlin, R. V. Nascimento, S. I. Watson, D. E. Fletcher, B. J. Koo (Savannah River Ecology Laboratory and Savannah River National Laboratory). **"Use of Illite Clay for In Situ Remediation of ^{137}Cs-Contaminated Water Bodies: Field Demonstration of Reduced Biological Uptake."** *Environmental Science and Technology*. Volume 40, Issue 14, July 15, 2006, Pages 4326-4528.

[59] Iwate, Miyagi, Fukushima, Ibaraki, Tochigi, Gunma, Saitama, and Chiba

[60] The Special Decontamination Area includes the entire area of the communities of Naraha, Tomioka, Okuma, Futaba, Namie, Katsurao and Iitate and some portions of the communities of Tamura, Minimi Soma, Kawamata, and Kawauchi.

[61] Fukushima City, Risk Management Office, presentation from March 7, 2013

[62] Public Involvement Policy of the US Environmental Protection Agency, May 2003, http://www.epa.gov/publicinvolvement/policy2003/finalpolicy.pdf

[63] Lessons Learned about Superfund Community Involvement, EPA Superfund Response Staff Tell How Public Involvement Has Helped Clean Up Sites, October 2009, http://www.epa.gov/superfund/programs/reforms/docs/leslrncomplete.pdf

[64] ***Better Decisions through Consultation and Collaboration, Conflict Prevention and Resolution Center, US EPA,*** http://www.epa.gov/adr/Better_Decisions.pdf

[65] Technical Assistance Grants, US EPA, http://www.epa.gov/superfund/community/tag/

[66] Community Advisory Group, US EPA, http://www.epa.gov/superfund/community/cag/

[67] Technical Assistance Services for Communities, US EPA, http://www.epa.gov/superfund/community/tasc/

[68] Superfund Job Training Initiative, US EPA, http://www.epa.gov/superfund/community/sfjti/

[69] General Community Involvement Toolkit Information, US EPA, http://www.epa.gov/superfund/community/toolkit.htm

In: Remediation of Lands After the Fukushima ... ISBN: 978-1-62948-846-2
Editor: Helen G. Turner

Chapter 2

THE JAPANESE NUCLEAR INCIDENT: TECHNICAL ASPECTS*

Jonathan Medalia

SUMMARY

Japan's nuclear incident has engendered much public and congressional concern about the possible impact of radiation on the Japanese public, as well as possible fallout on U.S. citizens. This report provides information on technical aspects of the nuclear incident, with reference to human health.

While some radioactive material from the Japanese incident may reach the United States, it appears most unlikely that this material will result in harmful levels of radiation. In traveling thousands of miles between the two countries, some radioactive material will decay, rain will wash some out of the air, and its concentration will diminish as it disperses.

Many atoms are stable; they remain in their current form indefinitely. Other atoms are unstable, or radioactive. They "decay" or "disintegrate," emitting energy through various forms of radiation. Each form has its own characteristics and potential for human health effects.

Nuclear reactors use uranium or mixed oxides (uranium oxide and plutonium oxide, or MOX) for fuel. Uranium and plutonium atoms fission, or split, releasing neutrons that cause additional fissions in a

---

* This is an edited, reformatted and augmented version of a Congressional Research Service publication, R41728, dated April 5, 2011.

chain reaction, and also releasing energy. A nuclear reactor's core consists of fuel rods made of uranium or MOX encased in zirconium, and neutron-absorbing control rods that are removed or inserted to start or stop the chain reaction. This assembly is placed underwater to carry off excess heat. The incident at the Fukushima Daiichi Nuclear Power Plant prevented water from circulating in the core of several reactors, causing water to evaporate and temperature to rise. High heat could melt the fuel rods and lead to a release of radioactive material into the air.

When uranium and plutonium fission, they split into smaller atoms that are highly radioactive and generate much heat; indeed, fuel rods that have just been removed from a reactor are much more radioactive, and hotter, than fuel rods before they have been inserted into a reactor. After fuel rods can no longer efficiently produce energy, they are considered "spent" and are placed in cooling pools of water for several years to keep them from overheating while the most radioactive materials decay. A concern about the spent fuel pool at reactor 4 is that it may have lost most or all of its water, yet it has more fuel rods than pools at the other five reactors, as it contains all the active fuel rods that were temporarily removed from the reactor core in November 2010 to permit plant maintenance in addition to spent fuel rods.

A nuclear reactor cannot explode like an atomic bomb because the concentration of the type of uranium or plutonium that fissions easily is too low to support a runaway chain reaction, and a nuclear weapon requires one of two configurations, neither of which is present in a reactor.

Some types of radiation have enough energy to knock electrons off atoms, creating "ions" that are electrically charged and highly reactive. Ionizing radiation is thus harmful to living cells. It strikes people constantly, but in doses low enough to have negligible effect. A concern about the reactor incident is that it will release radioactive materials that pose a danger to human health. For example, cesium-137 emits gamma rays powerful enough to penetrate the body and damage cells. Ingesting iodine-131 increases the risk of thyroid cancer. Potassium iodide tablets protect the thyroid, but there is no need to take them absent an expectation of ingesting iodine-131.

INTRODUCTION

The Japanese earthquake and tsunami of March 2011 caused extensive damage to the Fukushima Daiichi Nuclear Power Plant (NPP). This damage has released some radioactive materials, and there are widespread fears about the health effects of current and possible future releases. These fears, and public concern about radiation in general, have attracted the world's attention. This report presents scientific and technical aspects of these issues in order to provide a basis for understanding the risks associated with this event.

COULD HARMFUL LEVELS OF FALLOUT REACH THE UNITED STATES?[1]

To monitor radiation in the United States, the Environmental Protection Agency (EPA) operates RadNet, which "is a national network of monitoring stations that regularly collect air, precipitation, drinking water, and milk samples for analysis of radioactivity. The RadNet network, which has stations in each state, has been used to track environmental releases of radioactivity from nuclear weapons tests and nuclear accidents."[2] EPA has an online map of these stations,[3] and provides updates on the results of its air monitoring as it relates to the Japanese nuclear incident.[4]

Whether harmful levels of radioactive material from the incident reach the United States depends on many factors:

- Particle size: Tiny particles are more readily carried by the wind and can travel farther than large particles, which fall to Earth more rapidly.
- Wind patterns.
- Amount of material released: The more material released, the more likely some of it is to travel long distances.
- Melt vs. burn: If nuclear fuel rods (fresh or spent) melt and form a pool of very hot, highly radioactive liquid, that liquid might be contained by a containment structure. If it melts through that structure, it might contaminate groundwater. If the fuel rods burn, the fire would loft radioactive material into the air. The larger and hotter the fire, and the longer it burns, the more material would be injected into the air.
- Travel time: The longer radioactive material is in the air, the more of it will decay.
- Distance: The farther radioactive material travels, the greater the volume of air in which the material disperses, diluting it.
- Rain and snow: Precipitation washes some particles out of the air.

The first four of these factors depend on circumstances; the other three would reduce the amount of material reaching the United States under any circumstances.

According to U.S. nuclear authorities, the reactor incident does not appear to pose an immediate threat to the United States. On March 13, the Nuclear Regulatory Commission (NRC) stated, "Given the thousands of miles between the two countries [United States and Japan], Hawaii, Alaska, the U.S. Territories and the U.S. West Coast are not expected to experience any harmful levels of radioactivity."[5] On March 18, EPA and the Department of Energy stated that a monitoring station in Sacramento "today ... detected minuscule quantities of iodine isotopes and other radioactive particles that pose no health concern at the detected levels," and that between March 16 and 17, a detector in Washington State detected "trace amounts of Xenon-133, which is a radioactive noble gas produced during nuclear fission that poses no concern at the detected level."[6] In a briefing to the Nuclear Regulatory Commission on March 21, Bill Borchardt, NRC Executive Director for Operations, said, "natural background from things like ... rocks, sun, buildings, is 100,000 times more than any level that has been detected to date. We feel confident in our conclusion that there is no reason for concern in the United States regarding radioactive releases from Japan."[7] A press report of March 22 stated that equipment in Charlottesville, VA, detected radiation from the reactor incident, but that "health experts said that the plume's radiation had been diluted enormously in its journey of thousands of miles and that—at least for now, with concentrations so low—its presence will have no health consequences in the United States."[8]

It is useful to put these doses in perspective. Using the figure that natural sources provide 100,000 times the dose recorded in California and Washington state, it is possible to calculate a rough approximation of the dose from the Japanese incident, using the improbable assumption that the dose persists at the detected rate for an entire year. As discussed later, a report estimates that the average American receives a dose of 310 millirem (mrem) per year from natural sources. (Units of radiation dose are discussed under "Health Effects of Ionizing Radiation.") NRC requires its licensees to "limit maximum radiation exposure to individual members of the public" to 100 mrem per year. One-one-hundred-thousandth of 310 mrem per year is a dose of 0.00310 mrem per year. At that rate, it would take 32,258 years to accumulate a dose of 100 mrem; over a 70- year lifespan, the cumulative dose at this rate would amount to 0.22 mrem.

WHAT IS RADIATION?[9]

Many atoms are stable: they will remain in their current form indefinitely. Some atoms are unstable, or radioactive. They "decay" or "disintegrate," often transforming into atoms of a different element, such as through emission of radiation, which permits the atom to reach a more stable state.[10] The most common types of radiation emitted in decay, and their characteristics, are:

- Alpha particles are two protons plus two neutrons. They are electrically charged and massive by subatomic standards, and travel relatively slowly, so they lose energy quickly in matter. They travel only an inch in air, and are stopped by a sheet of paper or the dead outer layers of skin.
- Beta particles (an electron or positron[11]) are electrically charged, so are readily absorbed by matter, but are much less massive than alpha particles or neutrons. Depending on their energy, some are stopped by outer layers of skin, while others can penetrate several millimeters. They can travel up to several feet in air.
- Neutrons are typically emitted by heavy atoms like uranium and plutonium. They have no electrical charge and may be highly penetrating, depending on their speed. They can travel tens of meters in air; energetic neutrons can penetrate the body. They can be slowed down by hydrogen-containing material like water.
- Gamma rays are photons released during radioactive decay. Photons may be thought of as packets of electromagnetic energy; radio waves, light, and x-rays are less-energetic photons. Gamma ray energies vary widely. Those of medium to high energies are highly penetrating and can travel hundreds of meters in air. Stopping them requires a thick layer of a dense material like lead.

Several measurements are useful in discussing radioactivity. Radioactivity is measured in units of curies (Ci), where 1 Ci = 3.7×10^{10} disintegrations per second, or becquerels (Bq), where 1 Bq = 1 disintegration per second. (The curie is widely used in the United States; the Becquerel is more widely used internationally.) Specific activity—curies per gram—measures how radioactive a material is. Half-life is the time for half of the atoms in a mass of a particular type of radioactive material to decay. Specific activity is inversely related to half-life. For example, radioactive iodine-131 is intensely radioactive. It has a specific activity of 124,000 curies per gram and a half-life of 8 days; in 10 half-lives (80 days), 99.9% of the iodine-131 created at a given time will have decayed. In contrast, uranium-235 has a specific activity of 0.000002 curies per gram and a half-life of 700 million years; it would take 7 billion years (10 half-lives) for 99.9% of it to decay.[12] According to Richard Firestone, staff scientist, Lawrence Berkeley National Laboratory, uranium-235 emits so little radiation that "holding a piece in the hand would cause negligible radiation exposure."[13]

Energy released per decay is measured differently. A standard measure is the electron volt or, more commonly, thousands of electron volts (keV).[14] The penetrating power of gamma rays, and thus their threat to human health, increases as their energy increases.

Each radioactive atom, or "radionuclide," decays in a specific way. For example, when uranium-235 decays,[15] it emits gamma rays, most of which are of 186 keV (a low energy) or less, and alpha particles; cesium-137 emits gamma rays, virtually all of which are of 662 keV, a medium energy, and beta particles. Each radionuclide that emits gamma rays does so in a unique pattern, or "spectrum," of energies that is the primary characteristic used to identify many radionuclides.

RADIOACTIVITY AND NUCLEAR REACTORS[16]

Some heavy atoms, such as uranium-235 and plutonium-239, "fission" when struck by a neutron. In fission, an atom typically (1) splits into two lighter atoms, called "fission products"; (2) releases two or three neutrons; and (3) emits vast quantities of radiation. Fission products are often highly radioactive, such as cesium-137, iodine-131, and strontium-90.

Uranium-235 and plutonium-239 can support a nuclear chain reaction: to oversimplify, one neutron fissions one atom, which releases two neutrons that fission two atoms, releasing four neutrons that fission four atoms, and so on. Neutrons thus drive chain reactions; this is a key concept for understanding nuclear reactors. A supercritical mass supports an increasing rate of fission; fission diminishes in a subcritical mass; and fission proceeds at a constant rate in a critical mass. In an atomic bomb, a supercritical mass of uranium or plutonium supports a chain reaction that proceeds in a tiny fraction of a second, releasing vast quantities of energy. A nuclear reactor is designed to maintain a constant rate of fission. If fission proceeds too quickly, it gets out of control, in which case the fuel rods generate so much heat that they melt. When control rods are inserted into the reactor core, individual atoms continue to fission but the chain reaction stops. Control rods typically contain boron or cadmium because they are efficient neutron absorbers. (Because boron absorbs neutrons, it was added to cooling water in the Fukushima Daiichi NPP incident to prevent inadvertent criticality.) Fission that proceeds at the desired rate releases energy over several years from one load of fuel. The energy heats water to generate steam that spins turbines to generate electricity.

A nuclear reactor cannot explode like an atomic bomb because the fuels and configurations differ. In nature, uranium is 99.3% uranium-238 and 0.7% uranium-235. Only the latter is "fissile," that is, it will fission when struck by neutrons moving at relatively slow speeds. To make fuel for a bomb or a reactor, the fraction of uranium-235 must be increased through "enrichment."[17] An atomic bomb uses uranium enriched to about 90% uranium-235 ("highly enriched uranium," HEU), while nuclear reactor fuel is typically enriched to less than 5% ("low enriched uranium," LEU). LEU does not have enough uranium-235 to support a chain reaction of the sort found in an atomic bomb. In addition, a bomb must be configured in one of two ways to create a large enough mass to support a runaway chain reaction; reactors are arranged in an entirely different configuration.

A nuclear reactor uses pellets of LEU or mixed oxides (MOX, i.e., uranium oxide and plutonium oxide) for fuel. Fuel rods—thin zirconium tubes typically between 12 and 15 feet long—hold the fuel. According to one report,

> Zirconium is the metal of choice in this application because it absorbs relatively few of the neutrons produced in a fission reaction and because the metal is highly resistant to both heat and chemical corrosion.
>
> Low neutron absorption is vital to any structural material used in a nuclear reactor because large numbers of neutrons produced by the reaction must be free to interact simultaneously with all the nuclear fuel confined inside hundreds of fuel rods. This interaction sustains the necessary chain reaction throughout the reactor's core.[18]

Even with control rods fully inserted to halt the nuclear chain reaction, the radioactive decay of the fuel rods (primarily from fission products) generates heat, which must be dissipated. At the Fukushima Daiichi NPP, cooling was done by pumping cool water into the reactor. If the heat is not dissipated, the rods become so hot that they melt or burn. A fire would loft particles of radioactive material into the air. If fuel rods become too hot, their zirconium cladding may also react with water and produce hydrogen. The Fukushima Daiichi NPP primary containments used inert nitrogen gas to preclude hydrogen ignition. However, the operators had to vent the primary containment to relieve pressure, introducing hydrogen into the secondary containment, which is believed to have caused the explosions at reactor units 1-3.[19] This explains the urgency of the efforts to keep the fuel rods cool, and why the reactors suffered major damage when backup cooling systems failed.

In order to cool the fuel rods, personnel have been spraying huge amounts of seawater into the reactors and spent fuel pools. However, when seawater boils away from the heat of the fuel rods, it leaves behind large quantities of salt.

> The big question is how much of that salt is still mixed with water, and how much now forms a crust on the reactors' uranium fuel rods. Chemical crusts on uranium fuel rods have been a problem for years at nuclear plants.
>
> Crusts insulate the rods from the water and allow them to heat up. If the crusts are thick enough, they can block water from circulating between the fuel rods. As the rods heat up, their zirconium cladding can ignite, which may cause the uranium inside to melt and release radioactive material.[20]

To alleviate this problem, workers have begun using fresh water instead of seawater.[21]

As the fuel fissions in a reactor, the fraction of fission products in fuel rods increases. When the ratio of fission products to fissile material rises to the point at which a fuel rod can no longer efficiently maintain a chain reaction, it is referred to as spent fuel. "Spent" seems to imply that the fuel has been used up, and is therefore less dangerous, than fresh uranium fuel, but this is not necessarily the case. When fuel rods are first removed from a nuclear reactor, they have a high level of short-lived radionuclides, unlike new fuel rods, so they are intensely radioactive. This radioactivity generates intense heat, so spent fuel rods are placed in pools of water to cool them, typically for several years, until most of the short-lived radionuclides decay. The water also provides shielding against any radioactive release into the air, and the spent fuel pools have no hardened containment structure that would protect against radiation release. If a pool is drained, the fuel rods would heat up, melt, and perhaps burn. This possibility led to concern about the spent fuel rods at Fukushima Daiichi NPP reactor 4:

> The spent fuel pools can be even more dangerous than the active fuel rods, as they are not contained in thick steel containers like the reactor core. As they are exposed to air, the zirconium metal cladding on the rods can catch fire, and a deadly mix of radioactive elements can spew into the atmosphere....
>
> According to Tokyo Electric [Power Company]'s data, the spent fuel pool at the No. 4 reactor contains 548 fuel assemblies that were in use at the reactor until last November, when they were moved to the storage pool on the site. That means that the fuel rods were only recently taken out of active use and that their potential to burn and release radioactivity is higher than spent fuel in storage for a longer period.[22]

Another danger comes from the potential release of plutonium from the MOX fuel used at reactor 3. Even very small amounts of plutonium, if inhaled, can potentially cause lung cancer. This explains the concern about that reactor, as it is the only one that uses MOX fuel, although irradiation of uranium fuel also creates plutonium. Water is being pumped into the spent fuel pools at the Fukushima Daiichi NPP reactors as well to cool the fuel rods and prevent additional radiation release.

Health Effects of Ionizing Radiation[23]

Humans are continuously exposed to significant amounts of ionizing radiation from various naturally occurring and manmade sources. Because of its relatively high energy level, ionizing radiation is capable of producing significant biological change. Ionizing radiation gets its name from the fact that it causes ionization—ejection of electrons—when it interacts with atoms in the molecules that constitute cells and tissue. This process creates charged, often unstable, and highly reactive entities. The ensuing reactions may result in permanent molecular damage. Radiation disrupts cell division, which is why the most sensitive tissues are those in which cells frequently divide, such as skin, hair, bone marrow (where precursor cells give rise to new blood cells), and the cells that line the stomach and small intestine. Ionizing radiation may also damage DNA in chromosomes, resulting in mutations that are responsible for long-term effects such as the development of cancer.

Sources of Radiation Exposure

Naturally occurring sources of ionizing radiation to which all humans are exposed include cosmic radiation from outer space and terrestrial radiation from radioactive materials in rock deposits and soil. The Earth's atmosphere acts as a shield against cosmic radiation, so exposure levels increase with altitude (especially when flying). The most important source of terrestrial exposure is the inhalation of radon, which is produced by the radioactive decay of naturally occurring uranium.

In the United States, radiation exposure as a result of medical practice has increased significantly over the past 25 years as a result of the growing use of CT scans and nuclear medicine procedures to diagnose and treat disease. Other manmade sources of radiation account for a relatively small fraction of the U.S. population's total exposure. Those sources include consumer products (e.g., cigarettes, building materials, appliances); industrial, security, educational, and research activities, including nuclear power generation; and various types of occupational exposure.

Measuring Exposure: Absorbed Dose v. Equivalent Dose

Human exposure is measured by the amount of energy that ionizing radiation deposits in a unit mass of tissue. This is called the *absorbed dose*. The international unit for the absorbed dose is the gray (Gy), which replaced an earlier unit of dose, the rad (short for "radiation absorbed dose"). One gray equals 100 rad. The biological impact of ionizing radiation, however, depends not just on the absorbed dose (i.e., the amount of energy absorbed) but on the type of radiation. For example, an alpha particle is more damaging to biological tissue than a beta particle or gamma radiation because of its mass, electrical charge, and slow speed. Alpha particles lose their energy much more densely along the relatively short path they travel though biological tissue. Thus, 1 Gy of alpha radiation is more harmful than 1 Gy of beta or gamma radiation.

Table 1. Radiation Dose Levels

Dose, mSv	Dose, rem	Source	Comments
1/yr	0.1/yr	(2)	NRC requires its licensees to "limit maximum radiation exposure to individual members of the public" to this level.
6.2/yr	0.62/yr	(1)	Average U.S. individual's total effective radiation dose in 2006; half is from natural background and half is from medical uses and other human activities.
20	2	(7)	Federal Emergency Management Agency and Environmental Protection Agency recommend relocating the public from an area if the expected dose in the first year after a radiological incident is above this level.
50/yr	5/yr	(2)	NRC requires its licensees to "limit occupational radiation exposure to adults working with radioactive materials" to this level.
100	10	(6)	A National Research Council committee defines "low dose" of certain types of ionizing radiation, such as gamma rays, as this level or below.
0-250	0-25	(3)	For an "acute" (i.e., received over a short time) whole-body external dose of ionizing radiation, "No detectable clinical effects; small increase in risk of delayed cancer and genetic effects."
250	25	(4)	Japan raised the permitted dose for emergency workers at the Fukushima Daiichi NPP from 100 mSv/10 rem to this level.
500	50	(5)	For an acute whole-body external dose of ionizing radiation, "blood count changes."
1,000-	100-	(3)	For an acute whole-body external dose of ionizing radiation, "Minimal symptoms;
2,000	200		nausea and fatigue with possible vomiting; reduction in [certain white blood cells], with delayed recovery."
2,000-	200-	(3)	For an acute whole-body external dose of ionizing radiation, "Nausea and vomiting
3,000	300		on first day; following latent period of up to 2 weeks, symptoms (loss of appetite and general malaise) appear but are not severe; recovery likely in about 3 months unless complicated by previous poor health."

Table 1. (Continued)

Dose, mSv	Dose, rem	Source	Comments
3,200-	320-	(5)	Half the population exposed to an acute whole-body external dose of ionizing
3,600	360		radiation will die within 60 days despite receiving minimal supportive care.
3,500-	350-	(2)	NRC believes that half the population receiving this dose in a few hours or less
5,000	500		would die within 30 days.
8,000	800	(5)	100% mortality, despite best available treatment, for people receiving this external dose of whole-body ionizing radiation.

Sources: (1) National Council on Radiation Protection and Measurement, "Ionizing Radiation Exposure of the Population of the United States," report no. 160, 2009, p. 11. (2) U.S. Nuclear Regulatory Commission. "Fact Sheet on Biological Effects of Radiation," January 2011, http://www.nrc.gov/reading- and 10 CFR 20. (3) Dade Moeller, *Environmental Health,* revised edition, Cambridge, Harvard University Press, 1997, p. 250. (4) Keith Bradsher and Hiroko Tabuchi, "50 Workers Bravely Stay at Troubled Japan Reactors," *New York Times,* March 16, 2011. (5) Princeton University, Environmental Health and Safety. "Open Source Radiation Safety Training, Module 3: Biological Effects," http://web adapted from National Council on Radiation Protection and Measurements, Report No. 98, "Guidance on Radiation Received in Space Activities," Bethesda, MD, 1989. (6) National Research Council, Committee to Assess Health Risks from Exposure to Low Levels of Ionizing Radiation, "Health Risks from Exposure to Low Levels of Ionizing Radiation," BEIR [Biological Effects of Ionizing Radiation] VII Phase 2, p. 2, http://www.nap.edu/openbook.php?record_id=11340&page=1 and click on PDF Summary. (7) U.S. Environmental Protection Agency. Office of Radiation Programs. *Manual of Protective Action Guides and Protective Actions for Nuclear Incidents,* revised 1991 (second printing, May 1992), p. 4-4, http://www.epa.gov/radiation/docs/er/400-r-92-001.pdf, and Federal Emergency Management Agency, "Planning Guidance for Protection and Recovery Following Radiological Dispersal Device (RDD) and Improvised Nuclear Device (IND) Incidents," 73 *Federal Register* 45034, August 1, 2008.

Radiation scientists use another quantity, called *equivalent dose*, which allows them to measure all types of exposure on an equal basis. Equivalent dose is equal to the absorbed dose multiplied by a factor that takes into account the relative effectiveness of each type of radiation to cause harm. For beta particles and gamma radiation, the factor is set at 1; that is, the absorbed dose equals the equivalent dose. For alpha particles the factor is set at 20, which means that the equivalent dose is 20 times the absorbed dose. This reflects the fact that alpha radiation is more harmful than beta and gamma radiation. The international unit for the equivalent dose is the sievert (Sv). So, 1 Sv of alpha radiation to the lung would create the same risk of lung cancer as 1 Sv of beta radiation. The sievert is a large unit relative to common exposures, so the more common unit is the millisievert (mSv), which is one-thousandth of a sievert. The sievert replaced an earlier unit of equivalent dose, the rem, which is still widely used in the United States. One sievert = 100 rem; 1 mSv = 100 millirem (mrem).

The National Council on Radiation Protection and Measurement (NCRP) estimates that the *average annual equivalent dose* to an individual in the United States is 6.2 mSv (620 mrem).[24] Of that amount, 3.1 mSv (310 mrem) is from natural background sources, primarily inhalation of radon and its decay products, and 3.0 mSv (300 mrem) is from diagnostic and therapeutic medical procedures. The remaining 0.1 mSv (10 mrem) is from consumer products, industrial activities, and occupational exposure, among other sources. For comparison, the radiation dose from a jet airplane flight is 0.5 millirems (mrem) per hour in the air; from a chest x-ray, 6 mrem; and from living at an altitude of one mile, about 50 mrem/year.[25] Table 1 shows various doses and their health consequences or regulatory limits.

External v. Internal Exposure: Effective Dose

The health risks of ionizing radiation can occur as a result of both external and internal exposure. External exposure is almost exclusively from radioactive material that emits gamma radiation, which is very penetrating and, at higher energies, can only be stopped by a thick layer of lead or concrete. External sources of gamma radiation produce a whole-body exposure. Importantly, the level of exposure to gamma radiation falls off sharply with distance from the source. Cesium-137 (^{137}Cs), which has a half-life of 30 years, is the most common source of gamma radiation from nuclear weapons tests and reactor accidents.

Alpha and beta particles outside the body are typically not a source of external exposure. Alpha particles travel only a few centimeters through the air and cannot penetrate clothing or the outermost dead layer of skin. Beta particles, composed of electrons or positrons, can travel at most several feet through the air and penetrate to the live layer of skin causing burns (as happened to workers at Chernobyl). But they too are blocked by radiation suits.

Internal radiation exposure occurs through the inhalation of airborne radioactive material or the ingestion of contaminated food and drink. The potential for harm depends on the type and quantities of radioactive material taken in and the length of time they remain in the body. As already noted, isotopes that emit alpha particles present a greater hazard than those that emit beta particles and gamma radiation. In addition, the fate of the radioactive material depends on its chemical identity. For example, Strontium-90 (^{90}Sr), which is chemically similar to calcium and emits beta particles, accumulates in bone and can cause leukemia and bone cancer.

Iodine-131 (^{131}I), another beta emitter, tends to accumulate in the thyroid gland, where it is used in the synthesis of thyroid hormones. Beta radiation from iodine-131 damages the surrounding cells and increase the risk of non-malignant thyroid disease and thyroid cancer. Iodine-131 from radioactive fallout accumulates on grass and leafy crops and becomes concentrated in the milk of cows and goats that feed on the contaminated vegetation. Children who drink the contaminated milk are especially at risk because they are still growing and their thyroid glands are very active. However, iodine-131 has a half-life of only 8 days, so it decays relatively quickly on the ground, in the food chain, and in the body.

Iodine-131 posed the most important health risk following the incident at the Chernobyl nuclear power plant in 1986. According to the International Atomic Energy Agency:

> The main consequence of the Chernobyl accident is thyroid cancer in children, some of whom were not yet born at the time of the accident. Following the vapour [sic] explosion and fire at the Chernobyl reactor, radioactive iodine was released and spread in the surrounding area. Despite measures taken, children in southern Belarus and northern Ukraine, were exposed to radiation in the weeks following the accident, particularly by consuming milk from pastured cows and leafy vegetables that had been contaminated with radioactive iodine.[26]

Unlike whole-body external exposures, the exposure from ingested or inhaled radioactive material is often limited to certain parts of the body or even specific organs. Radiation scientists are able to calculate a whole-body equivalent dose, or *effective dose*, for partial-body exposures.

These amounts can be summed with external exposure to calculate a total dose.

Acute Health Effects v. Long-Term Cancer Risk

The health effects of ionizing radiation exposure depend on the total dose and dose rate. Radiation health experts distinguish between (1) acute, or short-term, effects such as radiation sickness that are associated with relatively high doses over a short period; and (2) long-term effects such as increased lifetime cancer risk that result from chronic exposure to low levels of radiation. Short-term health effects are typically seen in workers and others in close proximity to nuclear weapons tests and accidents, while the long-term cancer risks apply to the general population. Scientists calculate the cancer risk from radiation exposure using data from epidemiological and other studies, such as those following the health outcomes of the Japanese atomic bomb survivors. According to the International Commission on Radiological Protection (ICRP), the lifetime risk of contracting a fatal cancer from chronic exposure to low-level radiation exposure is 0.05 per sievert, or 1 in 20 per sievert (i.e., 1 in 2,000 per rem). The ICRP and NCRP both recommend an annual exposure limit of 1 mSv (100 mrem) for members of the general population. An individual that received that much annual exposure over a 70-year lifetime (a total of 70 mSv, or 7 rem) would, as a result, have an increased risk of cancer death of approximately 1 in 300.

Table 1 summarizes the health effects of exposure to various acute doses of ionizing radiation. For comparison, the table also includes the current exposure standards for the general public and workers, and the average background radiation exposure in the United States.

Potassium Iodide

There is considerable interest in potassium iodide (also referred to by its chemical formula, KI) tablets to protect against thyroid cancer. These tablets contain non-radioactive iodine-127, the same type used in iodized table salt, to saturate the thyroid with iodine. Once the thyroid is saturated, it cannot absorb more of any isotope of iodine, including iodine-131. As a result, potassium iodide tablets, taken shortly *before* exposure to iodine-131, offer protection from thyroid cancer. The protection is of limited duration, however, and potassium iodide protects only the thyroid only against radioactive iodine. It does not protect against any other radioactive material or against radiation in general. Nor is there value in taking potassium iodide as a precautionary measure unless iodine-131 is expected to be present. As the next section of this report discusses, the amount of radioactive material that has reached the United States from the Japanese nuclear reactor incident is minuscule. Accordingly, the website of the Centers for Disease Control and Prevention, accessed on March 22, said, "At this time, CDC does not recommend that people in the United States take KI or iodine supplements in response to the nuclear power plant explosions in Japan. You should only take KI on the advice of emergency management officials, public health officials, or your doctor. There are health risks associated with taking KI."[27] Further, "Some general side effects caused by KI may include intestinal upset, allergic reactions (possibly severe), rashes, and inflammation of the salivary glands."[28]

The Japanese Situation

Understanding dose and its health effects casts light on the Japanese situation. The (U.S.) Committee to Assess Health Risks from Exposure to Low Levels of Ionizing Radiation of the National Research Council reported on the health risks from a certain type of radiation that includes gamma rays and x-rays. It considered doses below about 100 mSv (10 rem) to be low doses. The committee found that many factors "make it difficult to characterize the effects of ionizing radiation at low levels," and that "at doses less than 40 times the average yearly background exposure (100 mSv), statistical limitations make it difficult to evaluate cancer risk in humans." To develop an estimate of risk, the committee constructed a "lifetime risk model [that] predicts that approximately 1 person in 100 would be expected to develop cancer (solid cancer or leukemia) from a dose of 0.1 Sv [10 rem] above background." For comparison, about 42% of the population will be diagnosed with cancer in their lifetimes.[29] At Fukushima Daiichi NPP,

> The workers are being asked to make escalating—and perhaps existential—sacrifices that so far are being only implicitly acknowledged: Japan's Health Ministry said Tuesday that it was raising the legal limit on the amount of radiation exposure to which each worker could be exposed, to 250 millisieverts from 100 millisieverts, five times the maximum exposure permitted for nuclear plant workers in the United States.
>
> The change means that workers can now remain on site longer, the ministry said. "It would be unthinkable to raise it further than that, considering the health of the workers," the health minister, Yoko Komiyama, said at a news conference. [30]

An acute dose of 250 mSv (25 rem) is the upper threshold at which dose is unlikely to cause noticeable health effects, but it increases the risk of cancer. Based on the National Research Council report, 25 of 1,000 people would be expected to develop solid cancers or leukemia as a result of receiving this dose. Workers exposed to this dose will probably not be allowed to be exposed to additional radiation above background for at least a year to give their bodies time to repair cell damage.

Beyond the Fukushima Daiichi NPP, the external doses reported fall far below the low-dose threshold of the U.S. Nuclear Regulatory Commission (NRC). Japan's Ministry of Education, Culture, Sports, Science and Technology reported dose readings from 80 monitoring stations between 25 and 60 km from the Fukushima Daiichi NPP.[31] On March 20, almost all the readings were less than 15 microsieverts per hour. (One millisievert = 1,000 microsieverts; 1 microsievert = 0.1 millirem.) At a rate of 15 microsieverts per hour, it would take 278 days to accumulate a dose of 10 rem. At the highest rate reported, 110 microsieverts per hour, it would take 38 days to accumulate that dose. Staying inside an uncontaminated building would reduce exposure considerably, and short-lived radionuclides like iodine-131 (half-life, 8 days) would decay significantly during a month or more, sharply reducing the dose they produce. On the other hand, a larger release of radionuclides would be expected to increase dose, and cesium-137 (half-life, 30 years) decays much more slowly than iodine-131, so it would contribute to dose for many decades.

Given the increase in thyroid cancer as a result of the Chernobyl disaster, a major concern in Japan is minimizing the risk of thyroid cancer. This is especially important for children. At Chernobyl, as noted earlier, ingestion of radioactive iodine-131 resulted mainly from drinking milk from cows that ate contaminated feed, and from eating leafy greens. Accordingly, Japanese authorities have tested spinach, other vegetables, and milk for iodine-131, and found elevated levels. In response, on March 23 Prime Minister Naoto Kan restricted the distribution and consumption of spinach, cabbage, broccoli, and other vegetables in Fukushima Prefecture, and restricted the distribution of fresh raw milk and parsley produced in Ibaraki Prefecture.[32] In addition, authorities have reportedly found traces of radioactive iodine in drinking water in Tokyo. On March 23,

> Ei Yoshida, head of water purification for the Tokyo water department, said ... that infants in Tokyo and surrounding areas should not drink tap water. He said iodine-131 had been detected in water samples at a level of 210 becquerels per liter, about a quart. The

> recommended limit for infants is 100 becquerels per liter. For adults, the recommended limit is 300 becquerels.... The Health Ministry said in a statement that it was unlikely that there would be negative consequences to infants who did drink the water, but that it should be avoided if possible and not be used to make infant formula.[33]

However, by March 24 the level was reported to be 79 becquerels per liter, and by March 27 had diminished to the point where two readings showed no radiation and one showed 27 becquerels per liter.[34]

APPENDIX. USEFUL LINKS[35]

U.S. Government

Centers for Disease Control and Prevention

"2011 Earthquake, Tsunami, and Radiation Release in Japan: Travel Information" http://wwwnc.cdc.gov/travel/content/id/2511.aspx

"CDC Situation Awareness - 2011 Japan Earthquake": includes some maps of Japan relevant to the nuclear incident

http://emergency.cdc.gov/situationawareness/japan/

"Radiation Emergencies": includes fact sheets on Japan: radiation and health http://emergency.cdc.gov/Radiation/?s_cid=emergency_001

Department of Agriculture

"USDA's Radiation Safety Questions and Answers": includes information on food imports from Japan

http://www.usda.gov/wps/portal/usda/usdahome?contentidonly=true&contentid= radiation_safety_qa.html

Department of Energy

"The Situation in Japan": includes statements, congressional testimony, photos, and detailed radiological assessments based on DOE's Aerial Measuring System http://blog.energy.gov/content/situation-japan

Department of State

"Embassy of the United States, Tokyo, Japan": includes information for Americans in Japan, updates on Tokyo water, availability of potassium iodide, travel warnings, press releases, etc. http://japan.usembassy.gov/

Environmental Protection Agency

"Japanese Nuclear Emergency: EPA's Radiation Air Monitoring": home page http://www.epa.gov/japan2011/

"Japanese Nuclear Emergency Radiation Monitoring: Daily Data Summary" http://www.epa.gov/japan2011/data-updates.html

"Radiation Doses in Perspective"
http://www.epa.gov/radiation/understand/perspective.html

Food and Drug Administration

"Radiation Safety": includes FDA actions to assess the situation in Japan and to ensure the safety of products imported from Japan
http://www.fda.gov/NewsEvents/PublicHealthFocus/ucm247403.htm

Lawrence Berkeley National Laboratory

"Exploring the Table of Isotopes": detailed data on isotopes of each element http://ie.lbl.gov/education/isotopes.htm

Los Alamos National Laboratory

"Ionizing Radiation—It's Everywhere," by Roger Echkardt, *Los Alamos Science,* no. 23, 1995: a detailed primer on ionizing radiation, written in a light style http://www.fas.org/sgp/othergov/doe/lanl/00326627.pdf

Nuclear Regulatory Commission

"NRC Actions on Japan's Emergency": statements and congressional testimony by commissioners, news releases, FAQs
http://www.nrc.gov/japan/japan-info.html

Japanese Sources

Government of Japan

"Countermeasures for Great East Japan Earthquake": messages from the Prime Minister, press conferences by the Chief Cabinet Secretary, links to ministries and agencies http://www.kantei.go.jp/foreign/incident/index.html

Japan Atomic Industrial Forum, Inc.

Home page: includes "Information Update on Fukushima Nuclear Power Station" with updates several times a day, such as on reactor status and major events, trend of radiation in the environment, other press releases http://www.jaif.or.jp/english/

Kyodo News

Home page: news feed, including reports on the nuclear incident http://english.kyodonews.jp/

Ministry of Education, Culture, Sports, Science and Technology

"Reading of Environmental Radioactivity Level": includes daily updates of radiation levels at many radiation detection posts

http://www.mext.go.jp/english/radioactivity_level/index.htm

Ministry of Health, Labour and Welfare

"Information about the Great East Japan Earthquake": announcements on radiation in food and water

http://www.mhlw.go.jp/english/topics/2011eq/index.html

NHK World

Home page of Japan Broadcasting Corporation: includes many news stories on the Fukushima Daiichi incident. Note: Members of the Board of Directors are appointed by the Prime Minister and approved by both Houses of the Diet.

http://www3.nhk.or.jp/daily/english/01_33.html

Nuclear and Industrial Safety Agency

"Countermeasures for Tohoku—Pacific Ocean Earthquake": includes updates of information on radiation levels

http://www.nisa.meti.go.jp/english/

Prime Minister of Japan and His Cabinet

"The Prime Minister in Action": includes messages, statements, and a blog by the Prime Minister http://www.kantei.go.jp/foreign/index-e.html

Tokyo Electric Power Company

Home page of the company that owns the Fukushima Daiichi nuclear power plant: includes press releases and monitoring data

http://www.tepco.co.jp/en/index-e.html

Other

American Nuclear Society

"Fukushima": Links to many news sources, updates on reactor status, an archive of updates and news

http://ansnuclearcafe.org/fukushima/

Areva

"The Fukushima Daiichi Incident," by Matthias Braun: provides a detailed description, with photos and diagrams, of plant design, accident progression, radiological releases, spent fuel pools, and sources of information

http://www.seyth.com/ressources/quake/AREVA-Document.pdf

Health Physics Society

"Fukushima Nuclear Plant Update": health aspects of radiation exposure in general and from the Fukushima Daiichi incident

http://www.hps.org/fukushima/

International Atomic Energy Agency

"Fukushima Nuclear Accident Update Log": daily detailed updates of the situation http://www.iaea.org/newscenter/news/tsunamiupdate01.html

New York Times

"Status of the Nuclear Reactors at the Fukushima Daiichi Power Plant": for each reactor, daily updates of the situation along with recent developments http://www.nytimes.com/interactive/2011/03/16/world/asia/reactors-status.html?ref=asia

"Assessing the Radiation Danger, Near and Far," by Joe Burgess et al., April 2, 2011: Combines data from multiple sources to show the potential danger at different distances and in various environments resulting from the Fukushima Daiichi incident http://www.nytimes.com/interactive/2011/04/02/world/asia/assessing-the-radiation-danger. html? ref=asia

Nuclear Energy Institute

"Japan Earthquake: Additional NEI Updates": includes daily updates and an archive of these updates, a timeline of events in Japan, and latest news and events http://www.nei.org/ newsandevents/information-on-the-japanese-earthquake-and-reactors-in-thatregion/japan-earthquake-additional-nei-updates/

Union of Concerned Scientists

"UCS Spotlight: Nuclear Reactor Crisis in Japan": daily press briefings on the incident, FAQs, the organization's position on nuclear power, and a report on NRC and nuclear power plant safety http://www.ucsusa.org/

World Health Organization

"FAQs: Japan Nuclear Concerns": Q&A from a United Nations Agency, including current risk, human exposure to ionizing radiation, WHO travel advice for Japan http://www.who.int/hac/crises/jpn/faqs/en/index.html

World Institute for Nuclear Security

"Emerging Lessons from the Fukushima Daiichi Accident: Maintaining Nuclear Security in a Complex Crisis": focuses on lessons learned from the incident for maintaining and improving security, as distinct from safety

http://wins.org/fileitem.aspx?id=249

World Nuclear Organization

"Nuclear Power in Japan": detailed background on many aspects of this topic including history; tables and descriptions of Japan's nuclear power reactors built, under construction, and planned; fuels used and spent fuel; and R&D. Does not deal with the Fukushima Daiichi incident. http://www.world-nuclear.org/info/inf79.html

"Fukushima Accident 2011": technical description of the accident, with diagrams http://www.world-nuclear.org/info/fukushima_accident_inf129.html

ACKNOWLEDGMENTS

The Nuclear Regulatory Commission provided technical comments on this report.

End Notes

[1] This section was written by Jonathan Medalia, Specialist in Nuclear Weapons Policy, Foreign Affairs, Defense, and Trade Division.

[2] U.S. Environmental Protection Agency. "RadNet—Tracking Environmental Radiation Nationwide," http://www.epa.gov/narel/radnet/.

[3] U.S. Environmental Protection Agency. "RadNet Map View," https://cdxnode64.epa.gov/radnet-public/showMap.do.

[4] U.S. Environmental Protection Agency. "Japanese Nuclear Emergency: EPA's Radiation Air Monitoring," http://www.epa.gov/japan2011/.

[5] U.S. Nuclear Regulatory Commission. "NRC Sees No Radiation at Harmful Levels Reaching U.S. from Damaged Japanese Nuclear Power Plants," press release no. 11-046, March 13, 2011, http://pbadupws.nrc.gov/docs/ML1107/ ML110720002.pdf.

[6] U.S. Department of Energy and Environmental Protection Agency. "Joint EPA/DOE Statement: Radiation Monitors Confirm That No Radiation Levels of Concern Have Reached the United States," press release, March 18, 2011, http://www.energy.gov/news/10190.htm.

[7] U.S. Nuclear Regulatory Commission. "Briefing on NRC Response to Recent Nuclear Events in Japan," public meeting, March 21, 2011, p. 13, http://www.nrc.gov/reading-rm/doc-collections/ commission/ tr/2011/20110321.pdf.

[8] William Broad, "Radiation over U.S. Is Harmless, Officials Say," *New York Times,* March 22, 2011, p. 6.

[9] This section was written by Jonathan Medalia, Specialist in Nuclear Weapons Policy, Foreign Affairs, Defense, and Trade Division.

[10] For descriptions of radiation, see Roger Eckhardt, "Ionizing Radiation—It's Everywhere," *Los Alamos Science,* no. 23, 1995, http://www.fas.org/sgp/othergov/doe/lanl/00326627.pdf, and U.S. Environmental Protection Agency, "Radiation: Ionizing and Non-Ionizing," http://www.epa.gov/radiation/understand/index.html.

[11] A positron is a positively-charged electron.

[12] For data on half-lives and other characteristics of radionuclides, see Lawrence Berkeley National Laboratory, "Exploring the Table of Isotopes," http://ie.lbl.gov/education/isotopes.htm, and U.S. Department of Energy. Office of Environmental Management. "Table B.1. Characteristics of important radionuclides," http://www.orau.org/ptp/ PTP%20Library/library/DOE/Misc/Table%20B_1_%20Characteristics%20of%20Important%20Radionuclides.htm.

[13] Personal communication, March 30, 2011.

[14] "An electron volt is a measure of energy. An electron volt is the kinetic energy gained by an electron passing through a potential difference of one volt." Fermi National Accelerator Laboratory, "How Big Is an Electron Volt?," http://www-bd.fnal.gov/ public/electronvolt.html.

[15] The number following the name of an element is the number of protons plus neutrons in the nucleus.

[16] This section was written by Jonathan Medalia, Specialist in Nuclear Weapons Policy, Foreign Affairs, Defense, and Trade Division, and Mark Holt, Specialist in Energy Policy, Resources, Science, and Industry Division. See also CRS Report R41694, *Fukushima Nuclear Crisis*, by Richard J. Campbell and Mark Holt. For the status of each reactor, see "Status of the Nuclear Reactors at the Fukushima Daiichi Power Plant," *New York Times,* http://www.nytimes.com/ interactive/2011/03/16/world/asia/reactors-status.html., and Japan, Nuclear and Industrial Safety Agency, http://www.nisa.meti.go.jp/english/.

[17] For information on the enrichment process, see U.S. Nuclear Regulatory Commission. "Fact Sheet on Uranium Enrichment," May 15, 2009, http://www.nrc.gov/reading-rm/doc-collections/fact-sheets/enrichment.html.

[18] "Zirconium: Covering for Fuel Rods," *New York Times*, June 9, 1995, http://www.nytimes.com/1995/06/09/ nyregion/zirconium-covering-for-fuel-rods.html.

[19] Information provided by Nuclear Regulatory Commission, personal communication, March 25, 2011.

[20] Keith Bradsher, "New Problems at Japanese Plant Subdue Optimism and Present a Risky Agenda," *New York Times,* March 24, 2011, p. 11.

[21] David Nakamura and Steven Mufson, "Japan Urges More to Evacuate," *Washington Post,* March 26, 2011, p. 1, and "Nuclear Energy—Crisis in Japan," *New York Times,* update of March 30, 2011.

[22] David Sanger, Matthew Wald, and Hiroko Tabuchi, "U.S. Sees 'Extremely High' Radiation Level at Plant, Focusing on Spent Fuel's Impact," *New York Times*, March 17, 2011, p. 13.

[23] This section was written by Jonathan Medalia, Specialist in Nuclear Weapons Policy, Foreign Affairs, Defense, and Trade Division, and C. Stephen Redhead, Specialist in Health Policy, Domestic Social Policy Division.

[24] National Council on Radiation Protection and Measurement, "Ionizing Radiation Exposure of the Population of the United States," report no. 160, 2009.

[25] American Nuclear Society, "Radiation Dose Chart," http://www.ans.org/pi/resources/dosechart/. This interactive chart permits the user to adjust values to find an approximation of his or her total annual dose.

[26] International Atomic Energy Agency, "Thyroid Cancer Effects in Children," staff report, August 2005, http://www.iaea.org/newscenter/features/chernobyl-15/thyroid.shtml.

[27] U.S. Department of Health and Human Services. Centers for Disease Control and Prevention. "Emergency Preparedness and Response: Radiation and Potassium Iodide (KI)," http://www.bt.cdc.gov/ radiation/japan/ki.asp.

[28] U.S. Department of Health and Human Services. Centers for Disease Control and Prevention. "Emergency Preparedness and Response: Potassium Iodide (KI)," http://emergency.cdc.gov/radiation/ki.asp#med.

[29] National Research Council. Committee to Assess Health Risks from Exposure to Low Levels of Ionizing Radiation. *Health Risks from Exposure to Low Levels of Ionizing Radiation,* Washington, National Academies Press, 2006, pp. 1, 2, 7, 8, http://www.nap.edu/openbook.php?record_id=11340&page=1, and click on "pdf summary."

[30] Keith Bradsher and Hiroko Tabuchi, "50 Workers Bravely Stay at Troubled Japan Reactors," *New York Times,* March 16, 2011.

[31] Japan. Ministry of Education, Sports, Culture, Science and Technology (MEXT), "Readings at Monitoring Post out of 20 Km Zone of Fukushima Dai-ichi NPP [Nuclear Power Plant]," news release, as of 19:00 March 20, 2011, http://www.mext.go.jp/component/english/__icsFiles/afieldfile/2011/03/20/1303972_2019.pdf.

[32] Japan. Policy Planning and Communication Division. Inspection and Safety Division. Department of Food Safety. "Restriction of Distribution and/or Consumption of Foods Concerned in Fukushima and Ibaraki Prefectures (in Relation to the Accident at Fukushima Nuclear Power Plant)," March 23, 2011, http://www.mhlw.go.jp/stf/houdou/2r98520000015wun-att/2r98520000015xym.pdf.

[33] David Jolly and Denise Grady, "Tokyo Says Radiation in Water Puts Infants at Risk," *New York Times*, March 23, 2010.

[34] David Jolly, "Radiation in Tokyo's Water Has Dropped, Japan Says," *New York Times,* March 24, 2011, and David Jolly, Hiroko Tabuchi, and Keith Bradsher, "High Radiation Found in Water at Japan Plant," *New York Times*, March 28, 2011, p. 11.

[35] This appendix was prepared by Jonathan Medalia, Specialist in Nuclear Weapons Policy, Foreign Affairs, Defense, and Trade Division.

INDEX

D

E

F

G

H

I

J

L

M

N

O

P

R

S

T

U

V

W

X

Z